KOMBUCHA
How-To and
What It's All About

By Alana Pascal with Lynne Van der Kar

$10.95 in the U.S.

Kombucha...How-To And What It's All About
By Alana Pascal with Lynne Van der Kar

Published by:
The Van der Kar Press
P. O. Box 189
Malibu, CA 90265-2855

Library of Congress Catalog Card Number 95-90065

ISBN-9645352-0-3

Printed in the United States of America
All Rights Reserved Including Foreign Rights

First Edition - 1995

This book may be purchased from the publisher. Please add $3.00 for postage and handling. For more than one book, add an additional $1.50 per book for shipping to the same address.

CREDITS:
Cover photo - Lynne Van der Kar
Cover photo collage - LiAnne Graves
Editor - Anina Maurier
Graphic Design - Jason D. McKean
Illustrations - "People"
Printed by Ben-Wal Printing, Pomona, CA

ACKNOWLEDGEMENTS...

This book would have never been accomplished without the help of the following people whose love, friendship and creative talents have contributed. Kudos and sincere thanks to LiAnne Graves for creating the photo collage on the cover and sharing with the world the talent that resides within her. Heartfelt thanks to Anina Maurier whose editing and patience transformed our rock into something that sparkles and shines. Thanks also to our friend who wants to be known as "People" for the illustrations and to Jason McKean for graphic design and cleaning up all the loose ends.

Without the love and vital support of the McKanns—Robin, Jane, Angelica, Malachi, Matthias, and Abbi—this project would not have happened. Also extended is appreciation and esteem for two amazing children, Sierra and Caleb, for their tolerance and personal grace.

And a special thanks to all those behind the scenes who gave us the love, unselfish help and support we needed to accomplish this task, among them Denise Ruben, Jerry L. Grannon and Dr. Arnie Berk, of the University of California, Los Angeles.

TABLE OF CONTENTS

FOREWORD

This book is written for all those who know nothing about Kombucha as well as those who have extensive experience with this living culture of yeast and bacteria. The phenomenon of Kombucha has swept the nation and is an indicator of the changes that are beginning to take place within our society concerning health and healing. The American medical system is being assaulted by an expanding grassroots movement which is seeking alternative treatment for conditions that are proving difficult for doctors to treat. Kombucha is currently the "hottest" factor in this growing trend. Extensive media attention has been focused on the culture and the beverage made with it. Some claim it to be an "elixir" or a "miracle cure" which has raised serious concerns within the alternative health community. We share those concerns.

This book contains basic care and feeding instructions for taking care of "mother" and "baby" cultures and tips on creating a stable environment for the production of a perfect beverage. How to recognize a serious problem, like contamination, is discussed as well as what are the proper dosages of the beverage for individual use. We attempt a thorough analysis of the chemical processes and the products involved in Kombucha brewing and have strived to not bore you with heavy scientific terms. However, some terminology could not be avoided, so a glossary is included.

More importantly, we have endeavored to include analysis of claims being made concerning the Kombucha beverage, and consequently we have stressed precautions regarding its use. If you can understand the product and how it might function, you will be better able to make a personal decision concerning your use of Kombucha rather than having to rely on hearsay and testimonials.

Responsible use of this "elixir" might provide health benefits to many individuals. Indiscriminate, irresponsible use could lead to its seizure and criminalization under the guise of protecting the public. After reading this book, please consider all the evidence and make as informed and educated a decision as is possible at this time. The responsibility is, after all is said and done, in each of our hands as individuals.

Cheers to your health and well-being.

PART 1

What It's All About

A BRIEF TALE OF FERMENTATION HISTORY

Approximately 10,000 years ago, somewhere in a valley in the Middle East, a Neolithic cavewife has made a disappointing discovery. Water has gotten into a large, open clay pot of cracked barley and has been left in this condition for many days. Because the big guy has not returned to their stone abode with a bleeding carcass recently, and the gathering season is well past, the family is compelled to gulp down the frothy liquid. Natural yeasts in the air attracted to the soaking barley have converted the gruel to a dark, bubbling brew. EUREKA!!! Beer! The naturally fermented grain has made a hearty robust high-protein food that has the added attraction of making the family feel rather "good."

Desiring to share their discovery throughout the cave neighborhood, they invite guests to gather around, dip their gourds and partake of the marvelous brew. With that hospitable gathering, an occupation and social pattern is instituted as the first "Cavern on the Green" draws the locals. More and more families decide that it is too much trouble to pack up the family after a night of imbibing and they set up camp. Villages are founded.

A few millennia moved toward the now, and at the Black

Auroch Inn somewhere in present day Iran or Iraq, Mrs. Tammuz, the proprietess and professional heiress of the legendary cavewife, is preparing the complimentary "between the Tigris and the Euphrates" classic breakfast of flat bread and figs for her customers. She has ground einkorn (a primitive form of wheat) to a flour. Her son, El-al, still in bed after a night of being frisky with the goddess' attendants at the temple celebration, has neglected to replenish the water containers at the village well, so the resourceful Mrs. Tammuz instead pours beer into the flour and mixes. Satisfied with her dough, she banks the hearth fire to warm up the frying stone. Shouts of rowdy patrons create a disturbance at the entrance of her establishment and she is forced to leave her fireside for a long time in order to bring peace to her inn. Returning to her hearth, Mrs. Tammuz discovers the flour and beer mixture has risen, the bowl has split, and the dough has oozed into the hearth and baked. VOILA!!! The first leavened bread is created. Restless customers are demanding to be fed and the ingenious Mrs. Tammuz passes off the calamity as the Specialty of the House. It is a BIG success.

Archeological evidence and anthropological speculation has arrived at such scenes pertaining to the introduction of fermentation as a food production process. Fermentation cultures (saccharomyces yeasts and lactobacilli bacteria) have been part of our dietary heritage for as long as civilization itself.

Our ancestral Julia Childs and Galloping Gourmets did not stop with fermenting grains. Soon in virtually every culture of the world individuals were allowing yeasts and bacterias to grow in such diverse food stuffs as yak milk, cabbage and soy beans. These fermented foods became a staple in traditional diets. Unfortunately, most children of immigrants to this country, in their desire to diminish their Old-World cultural identity, left out this wealth of fermented products from their diets in order to seem more modern and American.

Now, Kombucha has appeared on the American landscape. For the U.S., Kombucha is the current chapter of the fermentation history. But an intriguing mystery surrounds the beverage.

KOMBUCHA — THE MYSTERY...

The first and simplest step in any investigation into a mystery is to begin with the subject's name and its origin. In the case of Kombucha, this simple exercise has proved to be fraught with rumor and false leads.

The only real clue lies in the sound of the name. "Kombucha" or "Manchurian mushroom," the two names that have found popularity in the U.S., give the hint of an Oriental heritage.

Kombucha is by appearances a word of Japanese origin: *"Kombu"* being the Japanese word for kelp, an edible species of seaweed, and *"cha"* which means tea. Perhaps the seaweed was at one time brewed as a tea into which the culture was placed.

References to the beverage as "Manchurian" tea deepens the mystery. I first heard of this "secret rejuvenation tea from Manchuria" through an acquaintance when he displayed a slimy substance in a zip-lock bag and engaged me in a complicated story of high intrigue concerning its origin. This acquaintance called the soggy tortilla-appearing thing a "Manchurian mushroom" and claimed it was the direct descendant of the "original grandmother" which had been spirited into the United States taped to the body of a devotee at great personal risk. He mentioned the cost of bribes the smuggler was forced to pay Chinese officials totaling nearly $5,000. For $1500 his friend had obtained a "baby" and for $500 I, too, could partake of its magic. At that price I remained a skeptic. Months later a "baby" was left in my refrigerator, gratis. I became more curious.

No reference to this "ancient cure" occurs within the voluminous Chinese pharmacopoeia which is a composite of over 4,000 years worth of healing concoctions and includes the curative values of such diverse organic substances as earthworms, dried sea horses, praying mantis cocoons and stingray spines. How could this Kombucha have been overlooked? Today the beverage is reportedly sold in China by street vendors in the summer, offering the naturally carbonated drink as a thirst quencher with no curative or rejuvenative claims being made.

From Eastern Europe comes the name *"tea kvass."* *Kvass*, the Russian word for acid, is a traditional beverage rich in lactic acid and lactobacilli made from wheat or rye flour or black bread crusts, fruit and barley malt left to ferment in water. *Kvass* has a reputation of being of particular benefit to intestinal disturbances. When the Kombucha culture was introduced in its fermentation medium of black tea, the Russians referred to the beverage as *tea kvass.*

My personal interest in the mystery of Kombucha's origin was piqued when an elderly American woman, upon seeing the culture floating in its glass jar, claimed to recognize it as something her parents had kept in stoneware crocks in their cellar when she was a child. Her family, of Eastern European descent, then living in the Napa Valley of California, fermented wine vinegar with the gelatinous mass, calling the culture "mother of vinegar."

Is this "ancient cure" simply a vinegar culture that has been used by farmers and housewives throughout Eastern Europe, Russia and Asia for centuries? Or is it a symbiosis of two simple life forms, yeasts and bacteria, which in their mutually dependent processes can produce a product of remarkable human benefit? The recent discovery of a yeast (one of the same species which occurs in the Kombucha culture) which appears to have the ability to control the common cold virus, and that our greatest pharmaceutically manufactured life extender, penicillin, is a mold, allow adequate room for the possibility.

KOMBUCHA — HOW IT WORKS...

The Kombucha culture is a living organism composed of two symbiotic microorganisms, yeast and bacteria. It is not a mushroom. The yeast is of the Saccharomyces genus including the species ludwigii, apiculatus, cerevisiae, torula and a species called schizosaccharomyces pombe, which increases by cellular division rather than budding as the other listed species do. The bacteria are considered acetic bacteria including bacterium xylinium, bacterium gluconicum and acetobacter ketogenum. How they work together to produce the Kombucha beverage is most easily understood in terms of a factory.

This Kombucha factory (see illustration page 16) has two divisions, both doing essentially the same job but producing different products. The yeast cells produce acids and yeast, while the bacteria produce carbonation, ethanol and more acids including acetic (vinegar), lactic, gluconic and glucoronic acids.

15

**Gladys and Frank Sucrose
enter the Kombucha Factory**

The job is glycolysis—translated from the Greek as "the destruction of glucose," but rather than annihilation, what happens is a separation into even smaller units. In the "yeast division" of the factory, the 6 carbons of the glucose molecule are separated into a pair of 3 carbons. It is these three-carbon units (called pyruvic acid) which go on to produce energy.

This energy production causes the culture to propagate, growing to make a "baby," and to generate products.

Placing the Kombucha culture in the medium of sugar and tea is akin to the arrival of a factory into an area. The factory gathers and utilizes surrounding resources, altering its environment while producing goods.

The principal resource is the sugar in the medium. Sugar

provides an important commodity, energy, for the factory, energy in the form of employees and raw materials.

Sugar arrives at the personnel office of the factory in the form of a couple, Gladys and Frank Sucrose. Upon interviewing the couple, Gladys has displayed the skills to enter into the work place immediately, while Frank has to be placed into a training program to be "made over."

Sucrose is a disaccharide. A disaccharide is a two-molecule sugar. Sucrose is made of fructose (Frank) and glucose (Gladys). In both the yeast and bacteria divisions of the factory, Gladys (glucose) is the more valuable employee/asset.

Virtually all living organisms, including the yeasts and bacterias of the Kombucha culture, use glucose as the base unit of cellular energy production. The elaborate chemistry of digestion and assimilation is orchestrated for the reduction of all food (carbohydrates, proteins and fats) into this basic sugar. But with the Kombucha, the complicated digestive process is stripped down to its most basic steps. There is no alimentary canal to negotiate, no cellular gates padlocked without the right key as they are in the human body, just wide-open gates with welcome signs to the very simple cellular structure of the saccharomyces yeasts and acetic bacteria cells and one half of the sucrose couple can immediately enter the factory and begin to work.

Frank has to be referred to the company's image consultants (enzymes) to have a fashion make-over. Snip, snip. Shift. Glue, glue and parts of Frank's atomic anatomy are altered. The new redistributed weight means he needs support. But, in a few hours or, if he's a slow learner, a few days, he becomes a capable employee and now, he too, can enter the factory and begin work. (see illustration page 18)

**Fructose (Frank) is being altered by an enzyme
for utilization in the glycolytic pathway**

As with most processes where alterations and breaking down occur, there are byproducts. Of importance to Kombucha drinkers are some of these byproducts of our factory's energy production.

The factory analogy continues to describe the discharge of these byproducts. Out of the smokestacks drifts CO_2— carbon dioxide, out the drainage pipes comes ethanol, out the side doors go small quantities of organic acids, a number of them including citric, malic, oxalic and butyric acids. The primary product produced in this division of the factory is more yeast cells.

THE ACIDS...

Citric acid, like vitamin C, is an antiscorbutic factor, meaning it prevents scurvy. Chiefly found in lemons, limes, grapefruit, and oranges this acid is an essential Krebs cycle (also referrred to as the citric acid cycle) metabolite which initiates the energy producing processes necessary for life.

Malic acid has a reputation among herbalists and folk medicine advocates of enabling the liver to relax and discharge stored toxins perhaps as a result of its oxygen-carrying capability. Found in high concentration in apples and apricots, malic acid is one of the factors ascribed to the old saying "an apple a day keeps the doctor away." In addition, it has been recently reported that this acid, when used in conjunction with magnesium, is effective in relieving the muscular aches and pains associated with Chronic Fatigue Syndrome.

Oxalic acid is an important factor in the intercellular production of energy. However, oxalic acid is readily transformed to oxalate, an effective acidic chelating agent associated with stone formation within the kidney and gall bladder. One should not consume large quantities of this substance or it will bind up calcium and make it unabsorbable. But at the same time it occurs widely within the food we eat, and if the diet is balanced with adequate amounts of B6 (pyridoxine) and beta-carotenoids (found in green, yellow, orange and red vegetables), oxalates are less likely to precipitate into stones.

Butyric acid is a fatty acid most typically found in abundance in butter and cod liver oil and possesses a distinctive rancid odor. This acid is thought to have a role in protecting human cellular membranes. Most significantly it has been found to be of benefit in the treatment of candidiasis. This severe overgrowth of the yeast Candida albicans has wide-ranging and devastating health-robbing effects. Butyric acid has to be coupled with another fatty acid, caprylic acid, in order to be successful in stemming the growth of this insidious and hardy yeast commonly the cause of yeast infections and thrush. Butyric acid's less noble characteristic is noticeable when it is excreted by humans through sweat glands and interacts with bacteria on the body producing body odor.

YEAST???

Saccharomyces yeasts are used extensively in the food industry for the production of alcoholic beverages and the making of baker's and nutritional yeast. These yeasts have held respected places in traditional healing practices here in the U.S. as well as in Europe. In Kombucha, the yeasts precipitate to the bottom of the fermentation container imparting significant bioavailable nutrition into the medium.

Baker's yeast had a reputation as a remedy for indigestion, heartburn and constipation and was in widespread use in this country until the 1930's. At that time, a yeast company urged its customers to mix the cake yeast with fruit juice, or to spread it on crackers or bread. Unfortunately, this increased the amount of fermentation in the stomach and bowels causing autointoxication. Individuals were getting drunk from the breweries in their gut and toxic byproducts were entering the blood stream causing liver and kidney damage. The defamation of baker's yeast as a nutritional and medicinal substance arose from this unfortunate incident.

Baker's yeast was also reported as a treatment for acne and pimples. On the farm, breakfast is often served after the stock are cared for. My father told me stories of having to drink baker's yeast dissolved in warm water first thing in the morning before going out to do chores; Grandma reassured him that the ritual would keep his skin clear. When I reached my teens, my father tried to convince me of "Grandma's cure." He swore he came to like the taste. I found the thought of it too yucky to consider. He and his sister testified that neither had ever suffered the agony of teenage breakouts. I knew they were both too old to have accurate memories. I triumphed over the old ones' nonsense, slathered on mentholated cream out of cobalt-blue jars and suffered.

Research into old texts on this subject revealed that the "cure" was accomplished by the yeast "soothing a deranged liver" and oxidizing fats, perhaps by virtue of the B vitamins and the sulfur-containing amino acids, methionine and cysteine, found in yeast that had clogged the sebaceous glands (oil glands in the pores).

I still do not know about drinking baker's yeast. The practice is not even considered a topic of conversation among nutritionists, traditionally or alternatively trained. I never recommended it to my teenage children anticipating their "Oh Geez, Mom! Isn't it embarrassing enough we drink goat's milk?"-looks. It was an adventure I chose not to embark on in light of other more-appetizing alternatives.

NUTRITIONAL YEAST...

Nutritional yeast is still used extensively and recognized as a product of high nutritional value. This yeast product contains sixteen amino acids (including the eight essential aminos), fourteen minerals and seventeen vitamins. Its nutritional breakdown is impressive: 45 to 55 percent protein; 35 to 45 percent carbohydrates; 5 to 7 percent minerals; 0.5 to 2.5 percent fat. Nutritional yeast is traditionally regarded as a very beneficial and rather complete supplementary nutrient. We will consider in detail only the B vitamins, vitamin C, RNA and chromium. These very important and necessary substances for the proper functioning of our bodies are most significant when considering the claims of the Kombucha beverage's purported health benefits.

B VITAMINS

The B complex vitamins are active in providing the body with energy, essentially by converting carbohydrates into glucose. They are equally vital in the metabolism of fats and protein. They are necessary for the normal functioning of the nervous system and may be the single most important factor for the health of the nerves. Essential for maintaining muscle tone in the gastrointestinal tract and for the healthy condition of the hair, skin, eyes, mouth and liver, B vitamins are the foundation of most vitamin therapies. B vitamins are technically referred to as coenzymes and act as the principal ingredients of the vast majority of enzymes that the body produces for digestive and metabolic purposes. Nutritional yeast contains or produces all B vitamins except B12.

B1-thiamine—Thiamine is known as the "morale" vitamin because it is essential to a healthy nervous system. Thiamine provides an essential coenzyme to the energy-producing process of glycolysis. It is this process that provides us with energy.

B2-riboflavin—Riboflavin functions as part of a group of enzymes that are involved in the breakdown of fats. It is necessary for cellular respiration as an enzyme in the utilization of oxygen. Riboflavin also provides a component for a coenzyme necessary to energy production. As a point of interest, it is this B vitamin which colors the urine the characteristic bright yellow of B complex vitamin takers and does not denote overdosing.

B3-known by three names: niacin, nicotinic acid or niacinamide—Niacin is effective in improving circulation as a vasodilator and aids in reducing the cholesterol level of the blood. It is necessary to the proper functioning of the nervous system. It is also essential to the synthesis of sex hormones, cortisone (an adrenal hormone that is an anti-inflammatory), thyroxin (a thyroid hormone which regulates metabolism) and insulin (necessary for the transportation of glucose into the cells for energy production). Niacin recently became popular because of its ability to lower cholesterol. It has been used with some benefit in increasing joint mobility in arthritic joints due to its vasodilating attribute of increasing blood flow to the afflicted joints.

B5-pantothenic acid—This B vitamin is commonly called the "stress" vitamin and is the most widely dispersed of all vitamins throughout the body. The Greek root of its name, "*panto*," means everywhere. This vitamin is involved in virtually every human system: metabolic, digestive, hormonal and nervous. It stimulates the adrenal glands and increases production of cortisone and other adrenal hormones. Another noted benefit is its potential to retard the graying of hair.

B6-pyridoxine—Pyridoxine is yet another important co-enzyme in digestion and metabolism. Pyridoxine is known primarily for its ability to promote the efficient utilization of amino acids and essential fatty acids. B6 is also important in maintaining sodium/potassium levels necessary for proper kidney function and cellular fluid balance in the body. Because of its vital nature to a developing fetus, pyridoxine deficiency is conceivably a cause of morning sickness. To insure its correct development, a fetus takes priority over its mother for the use of her supply of pyridoxine. When the mother's body is depleted of this nutrient, her digestion becomes difficult. A pregnant woman's first warning of a potential pyridoxine deficiency could be morning sickness. Pyridoxine supplements have been found to be of benefit in most cases.

B15-pangamic acid—This nutrient was initially discovered by the Russians and supposedly included in all Russian military survival packets. Cold War politics being as they were, American nutritionists had to depend on foreign reports of the structure and benefits of this nutrient until our own research efforts caught up. Its primary benefit is as an oxygen deliverer, increasing the oxygen-carrying capacity of the blood by stimulating glucose oxidation. In short, it provides energy from otherwise insufficient sources. Of particular benefit to the heart muscle and glandular functions, it is said to increase the endurance of those who use it. Because of this quality it has found a place among athletes.

Biotin—This B vitamin is a coenzyme most effective in fatty acid metabolism. Without it, the body's fat production is impaired. Generally synthesized by the intestinal bacteria, its production can be seriously impaired by candidiasis. Biotin is another nutrient that has been associated with inhibiting the graying and the loss of hair.

Folic Acid—This B vitamin performs its basic role as a carbon carrier in the formation of hemoglobin, necessary for the production of red blood cells. Folic acid is so essential to blood building that to take an iron supplement without also supplementing folic acid is a virtual waste of effort. The fact is, in stubborn cases of iron-deficient anemia, giving more folic acid will usually raise the iron level when increasing iron by itself will have no effect. Folic acid recently received media attention for its ability to correct spinal bifida in afflicted fetuses. The effect of "mending" cleft palates had been previously documented and the same mechanism is employed in correcting the spinal deformity. This mechanism is associated with folic acid's importance in the formation of nucleic acids and the growth and reproduction of all body cells.

Choline and Inositol—These are two separate B vitamins whose functions are so intertwined that it is acceptable to group these two nutrients together. The pairing of choline and inositol is called lecithin. Lecithin is formed as an emulsifying agent making it easier for the body to utilize or discharge fat. Choline is an essential ingredient in the insulation of nerves as the principal constituent of myelin sheathing. Like plastic coating around a wire, this myelin sheathing is important in maintaining the proper firing of nerves. As a constituent in the calming neurotransmitter, acetylcholine, its value in soothing raw nerves is undisputed. Large quantities of inositol are found in the brain and throughout the central nervous system with particular benefit to brain cell nutrition. Remarkably, the body contains more inositol that any other vitamin except niacin. This is another of the B vitamins associated with preventing hair loss.

PABA-para-amino benzoic acid—This B vitamin stimulates the intestinal bacteria to produce folic acid. It is particularly active in the breakdown and utilization of proteins and in the formation of red blood cells. PABA is most widely known as a topically applied sunscreen. It protects the skin by absorbing portions of the ultraviolet rays known to cause burns and skin cancer. PABA combined with pantothenic acid (B5) is credited with the ability to restore hair that is turning gray and to prevent further graying.

AMINO ACIDS....

Hundreds of proteins which occur in the human body, involved in enzymatic activities, hormone production and protein structures of the musculature, are all made from smaller units. These smaller units are called amino acids. There are 24 amino acids, 8 of which are considered essential: methionine, leucine, isoleucine, valine, lysine, threonine, tryptophan and phenylalanine.

Of special interest to the Kombucha consumer is the presence of the sulfur-bound amino acids: methionine, cysteine, cystine and glutathione. Methionine is the essential amino acid from which cysteine and cystine are derived. These sulfur-containing aminos play significant roles as detoxifying agents specific to the liver, as well as providing essential elements of the SCoenzyme A molecule necessary to the production of energy. Methionine coupled with pyridoxine (B6) also aids in the maintenance of the pool of glutathione peroxidase, the powerful enzyme antioxidant which provides protection from the effects of alcohol, smoking and air pollution. The glutathione enzyme is severely depleted after large quantities of drugs or toxic chemicals are processed in the liver. Glutathione function is subject to the daily rhythms of physiology. Varying up to 30%, it picks up at night and peaks, then diminishes throughout the day. All sulfur-bound aminos will rid the body of heavy metals as well. Heavy metals are metals such as lead, aluminum, mercury and silver which will replace the naturally occurring metaloenzymes of a similar· valence or electrical charge and unless eliminated will cause a myriad of incomplete physiological functions ranging from DNA and chromosomal damage to nervous system and psychiatric conditions.

What you might notice as the "yeasty" smell in the Kombucha beverage is derived from the occurrence of these sulfur-bound aminos in the product.

RNA...

RNA, or ribonucleic acid, is a nucleic acid which transmits to each cell of the body instructions on how to perform. The body must be able to synhesize new RNA or cell degeneration begins. There is some controversy whether the RNA can pass through the stomach intact to impart its potential for cell regeneration but, nonetheless, the product has maintained a place within the supplement industry. Yeast is one of the best sources of RNA.

MINERALS...

The yeast produces numerous minerals including phosphorous, iron, sodium, sulfur, potassium and chromium.

Chromium sources are limited in the average diet. The yeast factory can contribute this necessary nutrient. Chromium is an essential element which the body uses most actively as an aid to blood sugar regulation. Chromium acts as the key which unlocks the door allowing the insulin molecule to piggyback glucose and pass into the cell. Referred to by nutritionists as a trace mineral because only small amounts are needed, it is considered by most nutrition authorities that Americans are deficient in this mineral because of their heavy dependence on highly refined foods which contain very little chromium and which also encourage urinary chromium reserve excretion. Recent nutritional reports indicate the mineral's ability to prevent hardening of the arteries and lower blood cholesterol and triglycerides. Chromium is also known to encourage the body to mobilize unwanted fat stores. The trace mineral has also been found to be essential in nucleic acid metabolism for rebuilding muscle tissue. Nutritional yeast is the most abundant source of this important mineral.

VITAMIN C...

Volumes and reams have been written about vitamin C. Dr. Linus Pauling, twice a Nobel prize-winning physicist, risked ridicule and professional ostracism for strongly advocating the use of high doses of this vitamin as a preventive measure against everything from the common cold to cancer. He based his theories on what he felt were rather compelling facts.

Humans share a physiological idiosyncrasy with primates and guinea pigs. We are the only mammals that do not internally manufacture a supply of vitamin C. Apparently, this biological diversion from other mammals occurred because our primitive diets had a very high natural vitamin C content. Our bodies evolved on the premise that we would graze and forage most of our waking hours eating a wide variety of fresh-off-the-vine foods that provided the needed vitamin C. Based on studies of primates (with whom we share a comparatively close physiological link) it was discovered that, in the wilds, primates consumed from 2,300 mg to 9,400 mg of vitamin C daily. Dr. Pauling reasoned that if we had not become "civilized," we would have continued on this browsing course and the vitamin C issue would not have become a critical one. Because of "civilizing" influences, humans began consuming a much narrower and nutritionally degraded range of foods through cultivation and storage practices. Added to this scenario, a greater majority of Americans in this century have acquired a penchant for highly refined and denatured foods. Finally, because our bodies do not store this vitamin, Dr. Pauling felt all these factors have left most of us severely depleted of this essential nutrient.

Vitamin C is a potent detoxifier, particularly of heavy metals as well as carbon monoxide and sulfur dioxide, which is common to air pollution. Vitamin C forms hydroxyproline and hydroxylysine, the protein fibers of collagen. It has been shown to kill viruses at doses of 10,000 mg and higher (taken in 1,000 mg doses spread out throughout the day) and is universally recognized as an immune system booster. Schizophrenia and depression have been associated with low vitamin C values. The adrenals need vitamin C to synthesize some of their hormones. Vitamin C has its most popular recognition as an antioxidant. It is because of this role as a free radical scavenger that vitamin C has been considered an important nutrient to increasing vitality and longevity.

THE BACTERIA DIVISION...

Back at the factory, the "bacteria division" has also swung into action, performing two jobs. Bacteria take up glucose and fructose (in its altered form of fructose 6 phosphate) for the fermentation process. Bacteria break down these molecules and, when adequate oxygen is present, fermentation will produce carbon dioxide and acetic acid. Acetic acid is the chemical term for vinegar.

The first job is involved in taking up some of the lactic acid and combining it to form lactobacilli. Lactobacilli are one type of bacteria found in the large intestine which encourage the proliferation of "good" flora. Balanced flora is essential to digestion and assimilation ensuring the proper and efficient uptake of nutrients out of the intestines. If these bacteria are disrupted, the colon's ability to synthesize the vitamins riboflavin, nicotinic acid, biotin, folic acid and vitamin K (the body's anti-hemorrhage factor which coagulates blood) would also be disrupted. Sadly, intestinal flora is severely upset by antibiotics and the ravages of stress. The inclusion of lactobacilli-rich foods has been a vital dietary practice among many cultural groups all across Europe and Asia. Sauerkraut, kim chee, miso, yogurt, kefir and kvass are some of the sources and are found as condiments, side dishes and beverages in traditional diets.

The bacteria break down glucose and fructose molecules by the fermentation process. The Kombucha fermentation process is best achieved when adequate oxygen is present to produce carbon dioxide and acetic acid. Along with a number of other organic acids, it is the conversion of the sugar medium towards the vinegar state that gives Kombucha its most potent health benefit. The beverage is beneficial, because it alters the pH of the liver, allowing it to detoxify itself of acid toxins.

THE BACTERIA-PRODUCED ACIDS...

Lactic acid occurs in three forms within the human body. Two forms of the acid are of particular importance concerning the Kombucha beverage. The first, a form of lactic acid called lactate (which is produced by fermentation) and found in the beverage, when delivered to the liver, will be rebuilt into starch and glycogen (a stored sugar the liver holds in reserve to be used in subsequent energy production). For every molecule of lactic acid recirculated in this manner 18 ATP molecules are created, making it a good source of energy. Dl-lactic acid is called fermentation lactic acid (found also in buttermilk and yogurt). In the stomach, it is extremely important for proper digestion and assimilation activities to occur. Without proper lactobacillus activity the ability to utilize food properly is seriously impaired and can become a detrimental factor to our health and well-being.

Gluconic acid is a significant ingredient in this milieu of the Kombucha-produced acids, especially considering the previously mentioned condition, candidiasis. Gluconic acid is a sugar acid produced by the bacterium gluconicum of the Kombucha culture. The intriguing aspect of this acid is that gluconic acid is biochemically named "an isomeric form of pentahydroxycaproic acid." If this acid is broken down (the five hydroxyl (OH^-) molecules removed) it forms caproic acid. Caproic acid can, and often does, go on to form caprylic acid. This means in theory that if the culture is composed of significant amounts of the aceter bacterium it could provide the necessary caprylic acid to compliment the butyric acid and could aid in remedying the pathogenic yeast overgrowth condition. *However, despite this possibility, Kombucha is not recommended for those who suffer from candidiasis. Because of the yeast inherent in the culture the possibility is too great that Kombucha could aggravate and seriously set back a candida patient. Consult a health practitioner about Kombucha before consuming the beverage if you suspect you suffer from this condition.*

Glucoronic acid is the liver's most potent and effective

detoxifying agent and can be readily transformed to hyaluronic acid which provides the foundation of our structural and skeletal systems. As a detoxifying agent, it is unsurpassed in aiding our survival in the modern world. When certain toxic products including phenol groups—derived from coal tar distillates or petroleum, and found as principal ingredients in pesticides, herbicides, resins and plastics—are ingested, they can be conjugated or "joined" in the liver by this acid and the offending substance is then eliminated efficiently out the kidneys.

In another capacity, glucoronic acid is transformed into glucoronides when exposed to free oxygen. These glucoronides make up hyaluronic acid. Hyaluronic acid is principal in the structure of mucopolysaccharides or glucosamines. Mucopolysaccharides are the structures associated with collagen, cartilage and synovial fluid. This is possibly the source of the claim that Kombucha can give the skin a healthy glow and reduce wrinkles by building collagen fibers. This might also account for its purported benefit to arthritis sufferers. A reduction of pain and swelling could be made possible by the replacement of the damaged and deficient joint fluids and cartilage with healthy versions.

THE pH STORY...

In 1958, Dr. Henry Jarvis, a country doctor from New England published *Folk Medicine in Vermont* expounding on the folk medicine wisdom he witnessed among and practiced with his patients. Vermonters, well known as independent souls and skeptical of anything newfangled, used such simple substances as apple cider vinegar, honey and kelp to treat virtually anything that could possibly "ail ye." It was through this book that thousands of Americans were first introduced to the possible benefits of drinking vinegar.

SO WHAT IS ACID-ALKALINE and the pH???

Understanding pH starts with water. Water is neutral, neither acid nor alkaline, registering 7.0 on the pH scale. It consists of two atoms of hydrogen (H) and one of oxygen (O)—H_2O. Water, however, tends to disassociate into units: OH^- and H^+. And with that simple combination magic begins.

In biochemical terms, water is considered a cohesive dynamic solution, but if we think of it as a ballet, we can break the scientific jargon down to a simple visual display.

Imagine that the principal dancers have taken the stage, OH^- (hydroxyl group, alkaline) the prima ballerina and H^+ (hydrogen ion, acid) her partner. The ensemble cast includes other ballerinas (OH^-) and male partners (H^+).

Act One. The scene is a simple village where life is easy and congenial. Many relationships are maintained in harmony and peace. Dancers leap, glide and pirouette about the stage, but our focus will be on the prima ballerina. Sometimes she comes together with another ballerina (forming hydrogen peroxide, HOOH) then leaving her, she moves on to pick up another male dancer (becoming H_2O again), and as the couple dance *pas de deux* to the corner of the stage, another male partner (H^+) joins the dancing pair to form hydronium (H_3O). The dance is being performed *allegro*, with "improv" performances by the many unattached ballerinas and dancers seeking, embracing and moving on, forming the same type of bonds and relationships, but constantly in motion. In this village, committed relationships do not exist.

33

Act Two. As the curtain rises, a handsome, charming stranger has entered stage right. His magnetic personality has compelled the prima ballerina to fall into his arms. Soon many of the ballerinas (OH⁻) rush to be similarly embraced. He gathers them to him and escorts them off stage, leaving the male dancers (H⁺) congregating at center stage as more and more of the ballerinas join the stranger and depart with him. The culprit has made the stage environment acidic. Many H⁺ dancers and fewer OH⁻ on stage mean there is no longer the neutrality of water. As fewer and fewer of the ballerinas find it difficult to resist the stranger's charms, the stage becomes an acid solution.

Act Three. The charming villain has departed and the village returns to normal. But the peace is short-lived as a carriage rolls onto center stage. The village occupants are hosts to another transformational presence. This time it is a remarkably beautiful and graceful nymph. The male dancers rush to be gathered in by her charms. She ruthlessly pulls them (H⁺) offstage as moths to a flame. Increasingly, the ballerinas (OH⁻) are abandoned to shift about listlessly as more and more of the ballerinas now crowd the stage. Elevated concentrations of OH⁻ molecules with fewer and fewer unattached males (H⁺) circulating among them and the stage has become alkaline.

Act Four. Her spell broken, the femme fatale has left the village; the ballerinas and the male dancers cavort once again, leaping, gliding and pirouetting as they perform the impartial Water Ballet once more. Peace and neutrality have returned to the village. The curtain comes down.

Acid pH is achieved when the solution contains more H⁺ (hydrogen ions) than water molecules (H_2O). Alkalinity happens when there are more OH⁻ (hydroxyl) molecules in the solution. On the pH scale, this means that the more H^+ ions in the solution, the more acidic the solution. Acidity is measured from .1 to 6.9 on the pH scale; .1 is the most acidic and 6.9 the least. 7 is the neutral point on the pH scale. Alkalinity starts at 7.1 and rises to the most alkaline at 14.

WHAT DOES THIS MEAN TO THE BODY?

The proper pH of body fluids and tissues is a major factor influencing health and well- being. When a pathogen invades a cell, its presence alters the pH of the cell rendering it damaged or, in virulent cases, dead. This is one of the ways that a disease or infection kills.

On a daily basis our bodies use or respond to compounds that are themselves acid or alkaline in nature, derived from the food we eat, the beverages we drink, the air we breathe and the emotional or psychological situations we find ourselves in. If the pH of the body fluids and tissues are not maintained effectively, enzyme patterns will "go awry." Waste materials will not be disposed of efficiently, resulting in toxicity, disease and eventually death. One of the many evidences of the remarkable intelligence of our bodies is the chemical activity that works diligently to prevent outer changes from disturbing the calm level of our inner environment. Let us take a quick look into this remarkable regulating system upon which our very lives depend.

THE BLOOD pH...

Our blood system, which regulates intercellular pH, is slightly alkaline—7.35 to 7.45. Slight variations outside of this range can result in coma or death. Even within the ranges of normal, there are conditions that are the result of being either "too alkaline" or "too acid."

Blood pH in the high normal levels—more alkaline—will tend to throw calcium out of solution. As the balance with calcium becomes more upset, the possibility that the endocrine system will become disturbed increases. Individuals in this condition complain about insomnia, tenderness of the joint areas, leg cramps, rheumatic pain and swelling, stiffness upon rising, and allergic-like reactions: runny nose producing clear to viscous mucus, sneezing, intestinal gas and irritability to a broader and broader range of foods, airborne pollens and dander. Hypoglycemia, poorly managed vegetarian diets, and a diet excessive in protein are three pathways to a "too alkaline" condition.

In contrast, an acidic condition is most dramatically encountered in individuals with diabetic blood sugar regulation problems. Diabetics do not properly metabolize their carbohydrates creating dangerous metabolic wastes (ketones) from fat and proteins. The result is clinically called acidosis and leads to diabetic coma and kidney failure.

Not as life threatening as acidosis, but more common, is a condition which stems from individuals with stress-related adrenal hormone production. In short, it is being "stressed out." These adrenal-acidic individuals initially complain of dry mouth, nausea, a marked gagging reflex, eye sensitivity to bright light and restlessness with an inability to quiet down after excitement. Prolonged stimulation causes excess secretion of adrenal-produced aldosterone which eventually depletes potassium reserves. Left unchecked, this leads to "adrenal exhaustion" which is an alkaline condition. With it comes fatigue, general debility, arthritic pain, immune deficiency and eventually physical and emotional collapse. Unfortunately for the sufferers of this syndrome, it is a condition not understood by the standard medical community as no definitive diagnosis of the complex symptoms can be established by lab tests. The treatment of this condition with its non-specific disease complaints is a common reason for individuals to seek alternative health care. Managing this condition is a lifestyle issue and cannot be cured with pills or surgery.

Many potentially acid-forming factors are encountered daily. The stressful nature of our lives, environmentally, emotionally and spiritually, and our common dietary habits come together to lead many into an unhealthy condition.

Eating the standard American diet (its acronym, SAD) is typical of one of these potentially stress-related factors. High in fat and simple carbohydrates, and distressingly deficient of grains, fresh vegetables and fruit, prevailing American dietary practices are disturbingly inadequate. Fruit, refined grains, coffee, tea, soft drinks, sugars, alcohol, nicotine and fried foods are all acid forming. Acid-forming food or drink produces an acid environment in the digestive system which eventually enters the blood system. This happens as the breakdown of these substances into glucose creates excessive amounts of metabolic wastes which in themselves are acidic. These are acetic acid and carbon dioxide. The same processes that have gone on in our factory go on in each and every cell of our bodies. The ridding of these wastes is generally accomplished through exercise and the increased respiration rate that accompanies physical exertion.

Compared to our grandparents, we participate in far less physical activity than our bodies require in order to keep us functioning properly. The prognosis for America is not good based on current trends. Changes have come swiftly to our culture, and together with our longer lifespans, we are not able to make genetic adaptations as fast as these changes have swept over us. In essence, we reside in genetically old-fashioned bodies living in radically artificial environments. It is not in our physical capacity to keep up.

The trick is to find a balance with your individual needs. Generally, a diet high in grains and fresh vegetables with the moderate consumption of meat, fish, beans, legumes and quality fats will accomplish good results. However, every person is different. I have yet to meet Mr. Normal or Ms. Average human being. Don't treat yourself as such. Each and every person has individual needs to be met based on age, genetics, physiological idiosyncrasies and lifestyle demands. Understanding your body is the first step in making sound dietary and lifestyle choices.

UNDERSTANDING YOUR BODY'S pH...

Alternating between alkaline and acid, the body has specific pH values for each stage of the digestion and elimination processes. Saliva in the mouth has a slightly alkaline 7.1 pH necessary for the initial digestion process to begin. (Saliva contains amylase and ptyalin, alkaline enzymes which initiate carbohydrate digestion.)

The stomach fluctuates from .9 to 2.0 (very acid) and sometimes the pH will rise to a less acidic state. If the acid level becomes more alkaline (3.0 or higher), digestion problems can arise.

In the small intestine, the pancreas and gall bladder work together to create an alkaline environment so that the serious work of digestion can take place. Pancreatic enzymes perform best at pH values of 7.1 to 8.23.

On to the elimination phase of digestion. In the large intestine, the pH once again moves to an acid state of 6.5 to 6.8. The kidneys also discharge urine at an acid level of 6.8. (Usually, the first urine in the morning will be the most acid. Generally the urine pH will alkalize slightly during the day.)

As you can see, pH values are vacillating but never randomly. The values that have been listed here are for healthy individuals. Testing the saliva and urine with pH strips is a simple way to determine if an imbalance exists.

TESTING YOUR pH...

The pH of the saliva and the first morning's urine can be good indicators of the health of your digestive system.

Orally—one half hour before or after a meal—simply place a pH strip in your mouth. It only takes a second. Remove it. The strip should register between 7 to 8 pH. If the pH runs on the acid side, this indicates problems with buffering systems and could mean a chronically acid blood condition, the prelude to a very serious over-alkaline condition because acid blood tends to throw calcium out of solution.

Urine—placing a strip under the midstream flow of the day's first urination—should give you a reading in the 6 to 7 range on the acid side. If alkaline, the body is not effectively disposing of toxic compounds that should be being excreted.

HOW THE BODY HANDLES pH PROBLEMS...

There are two major pathways for ridding the body of excess acid: excretion (or elimination) and buffers.

Excretion includes multiple ways of eliminating toxic

acids. Breathing is the first level of defense. Shallow breathing does not allow enough carbon dioxide to be effectively exhaled, causing a "backup" within the blood that will raise the blood acid level. This situation leads to racing heartbeat, shortness of breath and agitated movement and behavior. Taking three deep breaths—inhaling through the nose and exhaling out through the mouth—can alkalize the blood pH significantly and calm the nerves, the heartbeat and the mind.

Offending substances can be detoxified by dilution, a common cause of edema. Edema happens as the body uses water to thin or dilute toxic compounds which tend to be acid. The fluid often accumulates in the hands, feet, ankles and knees causing painful swollen joints. In some instances, the use of a diuretic will reduce the fluids by flushing toxic fluids out the kidneys.

It is the second pathway that most interests us when speaking of the Kombucha beverage. This pathway is called the buffering system.

Perhaps Kombucha's greatest health benefit comes from its ability to cover all the pH bases, delivering acid to stimulate the detoxification of the liver while providing potent buffering agents to the blood stream and eliminating toxic compounds out through the kidneys with a mild diuretic effect. How?

The acetic acid of the Kombucha beverage and its very acid pH of 2 to 3 passes through the gut, enters the blood system and is immediately taken up by the portal vein into the liver. This acid situation, and enhanced by malic, citric and glucoronic acids and the sulfur-containing aminos of the yeast, causes the liver to discharge, dumping toxins into the blood which move out via the vena cava into the general circulatory system. Once in the bloodstream either the substances are being shuttled off with the help of the glucoronides to the kidneys for elimination, or the carbonation has stimulated the alkalizing agents that can buffer those substances, rendering them null and then passing them out through the kidneys.

Some of the acids that have been mentioned to be contained in the Kombucha beverage, including acetic, glucoronic and citric acids, are weak organic acids and act as part of a primary natural buffer system necessary to safeguard against potentially damaging conditions by maintaining the proper pH for cellular function. Buffers inactivate situations either too acid or too alkaline. Kombucha contains both acid and alkaline buffers.

However, the champagne-like bubbles of the Kombucha beverage are the product of carbonation. Carbonation is carbon dioxide and it is very acidic. Remarkably, this acidic carbon dioxide provides the most dramatically effective acid buffering system—rendering substances alkaline or neutral—available to the human body.

Confused? Actually, once again a complicated biochemical process can best be explained in the terms of a dancing team.

BLOOD BUFFERS...

Let us go back to our factory. You will remember that carbon dioxide (CO_2) is one of the factory's byproducts. Being a very attractive acid molecule, carbon dioxide (CO_2) is spry and ready to dance after being spewed out the factory smokestack. As he drifts down into the crowded active neighborhood of the tea medium he attracts and is readily joined by the dancing pair of a water molecule (H_2O or OH^- and H^+), they form the Carbonic Acid (H_2CO_3) Dance Team. Ah, but

41

their bond is weak, and the H^+ ion is feeling unfulfilled. Being encouraged by friends, he strikes out and links up with a free roaming OH^- molecule and becomes part of the Water Ballet (H_2O) once again. After the departure of their acidic colleague, the Carbonic Acid Dance Team has been transformed into bicarbonate (HCO_3) and has become alkaline by virtue of the OH^-(alkaline) molecule that remained in the dance team. But the bicarbonate duo is desperate to conjugate (join) with an ionized salt to form a much-sought-after touring troupe.

A number of potential dancers arrive at the audition. Potassium has great legs but has a very tight schedule as he is needed in too many places at once. Calcium hasn't the grace the pair thinks necessary. Magnesium couldn't get the dance steps right for the performances the Troupe are scheduled to exhibit. So, the pair decide on the very lithesome and flexible sodium molecule and sign the contract. Now the three have formed the remarkably potent Traveling Buffering System Troupe of sodium and bicarbonate ($NaHCO_3$) and together they embark on an extensive tour along the highway of the circulatory system. They find eager audiences and perform to rave reviews.

Already-formed carbonic acid as well as carbon dioxide (carbonation) are ingested when one drinks the Kombucha beverage. Passing through the acid environment of the stomach, these acids are readily absorbed into the blood stream. Once there, carbon dioxide combines with OH^- molecules in the ever-shifting dynamics of the water molecules and with highly alkaline ionized salts (magnesium, calcium, potassium and sodium) found within the blood stream. All three factors: CO_2, OH and mineral salts come together to form bicarbonates—the most potent of the body's alkalizing agents. Bicarbonates neutralize acid compounds that have found their way into the blood system, be they the carbon dioxide of basic respiration or the acid-forming foods and beverages we come in contact with which tend to be acidic.

Blood buffers are the elements that keep the pH from fluctuating from one extreme to the other and resist changes in the concentration of H^+ (acid). Although the body, generally speaking, has adequate amounts of carbonic acid at its disposal without the ingestion of the carbonated acid liquid of Kombucha, it is the ability of Kombucha to deliver chemicals that can cause slight variations within the normal range that makes a difference.

Proper pH is fundamental to energy production and the elimination of metabolic wastes. If the pH is not appropriate, a host of enzyme and hormonal activities are either slowed down, speeded up or nullified. This sets the stage for disease and chronic degenerative conditions that prove to be difficult to treat. Many alternative therapies address this pH issue. Drinking vinegar or Kombucha, ozone therapy, hydrogen peroxide therapy and adequate exercise can all have dramatic effects on maintaining or achieving proper pH levels which can mean the difference between a sluggish, depressed state of being and a *joie de vivre*.

The human body has a particular organ, the liver, that is the ultimate determining factor as to the quality of your life, emotionally and physically. It is slavishly codependent in its relationship to you, constantly working, suffering and striving to improve your quality of existence. And, chances are, many have no concept or appreciation of its abilities until it is much too late.

THE LIVER...

In ancient times, the liver was considered the seat of the soul and therefore the most important organ of the body. In traditional Chinese Medicine it is designated as the dispenser of "*chi,*" the life force. (In Western terms, it can be explained in relation to the thyroid hormone, thyroxin, being used by the liver to generate metabolism.) The liver is the second largest organ of the body (second only to the skin) and, undoubtedly, the most abused. Responsible for over 100 known functions, the liver is a marvel of chemical precision with diverse abilities: master chemist, supply office, fuel storage, housekeeper, and poison control center.

The master chemist portion of the liver metabolizes proteins, fats and carbohydrates providing energy and nutrients for the other systems of the body. It creates bile to help emulsify fats making them easier to break down into their fatty-acid components, works to maintain electrolyte (magnesium/calcium and potassium/sodium) levels, is involved in the formation and breakdown of blood and helps to maintain water balance by producing serum proteins.

As a supply office, the liver will provide extra blood on demand in case of a critical situation. It also stores fat-soluble vitamins A, D, E and K.

As a fuel storage facility, it stores B vitamins, minerals and sugars. The latter are stored as a reserve sugar called glycogen which is meant to be released when blood stores of available glucose become low due to exercise, stress or delayed eating.

Functioning as the major detoxifying organ of the body, the liver is responsible for filtering the blood and removing harmful bacteria and chemicals including the breakdown and elimination of excess hormones. This function of our liver has been dramatically overburdened within the last two generations because of the impact of our polluting technologies and lifestyle habits: unending stress, poor diet, alcohol and drugs. It is by this means that we have wreaked havoc on this amazing organ and contributed more to our "dis-ease" than ever before in our species' sojourn upon this planet.

We have been blessed with a remarkably resilient and faithful servant, but how long it can stand this abuse is a critical question as your faithful loyal friend, protector and lover, the liver, puts up with it all. Believe me, your liver LOVES you. It works so hard to make things right and good and comfortable for you. But after a time, subject to insult after insult, the liver can become "deranged."

HOW WOULD YOU KNOW THE LIVER IS "DERANGED?"

Chinese medicine has a very elaborate list of liver symptoms. Western-trained doctors will be in a swoon trying to make "clinical sense" of this list. Chinese medicine is a system that requires of Westerners a repatterning of the brain. The first step is surrender. The second step is observation.

The liver expresses itself through the eyes. Red, inflamed conjuctiva and sclera, watery, painful, feeling of sand in the eyes, blurred misty vision, film over the eyeballs and guck in the corners are all signs of liver imbalance. The amount of life and vitality that the eyes reflect is a very good indication of liver health. A dull, listless, unfocused or tired look signifies problems. Peace, compassion, love, mirth and joy are states of being that the eyes convey when the liver is healthy.

If the liver is not "watering the yin" or not able to disperse nutrients properly, ligaments and tendons become tight.

There is very limited flexibility. Knees, shoulders, hips and other joints do not articulate well. Aching joints upon waking in the morning and arthritis also fall into this category.

The nails reflect the quality of the liver. Split, flaking, ridged, pale or brittle nails indicate liver disharmony.

Painful swelling of the breasts and pain on the left and right sides of the body in the lower rib area indicate a potential liver problem. A bruised or full tightness of the area just under the lower curve of the bottom ribs on the right side of the body is a good indication of liver problems by Western criteria.

Headaches, and in particular a "liver full" headache (this type of headache involves sharp excruciating pain, nausea and vomiting) vascular, unilateral headaches called migraines, flushed face, bright "apples" in the cheeks, hot flushed feeling, and hot flashes are "liver excess" symptoms. Acne, psoriasis and eczema, clear to white mucous discharge from the nose typical of allergies and hayfever, also ringing in the ears, middle ear infection (otitis media), and dizzy spells are also linked to the liver.

Bitter taste in the mouth, dry mouth, a craving for sour foods: green apples, lemons or vinegars, indicate an irritated liver. According to folk and Chinese medicine, sour will soothe the liver and appears to be the reason for the craving.

Waking between 1 and 3 a.m., sometimes sweaty and agitated, and unable to return to sleep for sometime, is a form of insomnia typical of liver dysfunction. Centuries ago, the Chinese developed a "celestial clock" based on when the flow of energy is at its peak for each organ system. According to this clock, and Western clinical research, the liver is most active at these early morning hours. Waking at this time is recognized by some physicians as a typical symptom of stress and Chronic Fatigue Syndrome (CFS). In CFS, the liver is inflamed. Because of this, the liver responds to normal daily rhythms of function with excess activity.

The liver *would like to* wind down from 3 p.m. until midnight. Bile production is at its lowest rate in the late afternoon into the evening. This is the root of the practice in rural and traditional areas of the world that the main meal of the day is eaten midday with a light meal in the evening and signifies an innate wisdom that has been lost as the Industrial Age has progressed. At near midnight, bile production is increased and in cases where the liver is inflamed, the patient will be disturbed by this increase of liver activity causing the patient to wake and be unable to fall back to sleep immediately. The "liver- friendly" practice of retiring early and rising early is reflected in the liver's physiological rhythms. Liver repair is only successfully accomplished when the patient sleeps. The need for good adequate rest for healing becomes apparent. Late evening meals and keeping late hours regularly burden the body dramatically, especially as the body ages.

Fatigue, irritability, lack of determination, being easily upset, short tempered, feeling nervous sensitivity and attention to trivial matters are all psychological "states of being" associated with an aggravated or depleted liver. The emotion typically associated with the liver is anger, flashing, aggressive outbursts or just a plain mad-at-the world attitude. Dreaming of war, fights, fighting and destruction are liver symptoms that reveal themselves while we sleep.

Fear of exercise, making excuses or not desiring to commit to an exercise routine or not wanting to be involved in physical activity, depleted sexual desire, white, mucousy, fishy smelling discharge of the vagina, whitish discharge of the penis can also be indicative of liver problems. Menstrual activity is greatly influenced by the health of the liver which means that menstrual irregularities including excessive bleeding, cramps, light menses or the lack of menses all have a liver connection.

This is a partial list of symptoms associated with liver pathology. In Chinese medicine, there are 6 main possibilities of pathology or disharmony. Some symptoms from all the possibilities have been included. Not a single symptom mentioned above is considered a normal occurrence for a healthy body.

Healing the liver means reducing toxic encounters in diet, emotions, environment and lifestyle.

Because of its role in pH regulation and its combination of liver-detoxifying acids, Kombucha can improve the liver's function and its ability to heal. These attributes are the most valuable in assessing the benefits of drinking the Kombucha beverage.

⚜ PART 2 ⚜
Care And Feeding

WHAT YOU NEED

- *one Kombucha culture*
- *a stainless steel, Corningware or Pyrex 4-quart saucepan*
- *a wide-mouthed glass jar, Pyrex bowl or stoneware ceramic bowl that will hold up to 4 quarts.*
- *a clean air-permeable cloth, i.e. a cotton dishtowel, two to three thicknesses of cheesecloth or other cloth cover that will allow the culture to breathe without allowing in dirt or insects.*
- *a rubber band, string, twisties or tape which will secure the cloth around the jar or bowl's neck.*
- *3 quarts pure filtered water*
- *1 cup white sugar*
- *5 tea bags (2 grams each-individual size) of black tea or a combination of black and green tea. The more caffeine content, the better the quality of the beverage that is produced.*
- *4 ounces of reserve Kombucha beverage from a previous brewing or 1 ounce of white or apple cider vinegar.*

Place 3 quarts of filtered pure water into the saucepan. Dissolve the sugar in the water as it comes to a boil and boil for five minutes. Take the saucepan off the stove and place the tea bags in the sugar solution. Cover and let steep for 10 to 15 minutes. Remove the tea bags and allow to cool to room temperature. Pour the solution into the glass container or bowl that is to be your brewery. Pour the reserve Kombucha (or vinegar) into the liquid. Place the Kombucha culture in the liquid. Cover the container with the cloth. Secure with a rubber band or string. Place it in a warm, dark, quiet place. In approximately one week, your Kombucha beverage should be ready.

A VARIATION...

Some individuals have found green tea (to "soften" the tea medium) a viable alternative, substituting half the black tea with green tea.

BREWING KOMBUCHA

The most dependable way to check a brew's readiness is to use pH strips in order to test for proper acidity. After 6 to 7 days, check the brew by inserting the pH strip into the liquid. When the strip registers a pH of 3, the beverage is ready. (see resources)

With clean hands, withdraw the culture from the container and place on a clean plate. If the "baby" is attached to its "mother," seperate them gently. You may re-use the cultures immediately or you can store them for future use. Using a plastic strainer or cheesecloth, strain the beverage into a holding container, a glass or ceramic pitcher or glass jar. Take one-half cup of the finished beverage to hold in reserve for the next batch. If you are going to make another batch immediately, leave the yeast sediment on the bottom of the fermentation container. This sediment should be rinsed from the container every few batches. Leaving the yeast sediment in the bottom of the container will enhance the beverage's benefit by increasing the yeast-derived nutrients. From here, just repeat the process.

Although Kombucha is a vinegar culture, the idea for making the Kombucha beverage is to harvest it before it reaches a full vinegar condition. Under good conditions, approximately one week into the fermentation process is the time the beverage is the most palatable. If allowed to sit longer, it becomes too vinegary for most individuals.

The development of particular characteristics of the beverage are dependent on the nature of the medium, the air temperature of the room in which the tea is "brewing," and the availability of oxygen.

DRINKING KOMBUCHA...

Begin your Kombucha experience by drinking small amounts of the beverage—2 to 4 ounces—in the morning before eating breakfast for 10 days. This test period should determine how your body responds to the beverage. This step is important for individuals who have unknown yeast allergies or undetected candidiasis. If an individual develops:

- *nausea*
- *a rash across the upper body*
- *itching and flushing of the face, particularly around the nose and across the forehead*
- *intestinal gas and cottage cheese-appearing matter in the stool*
- *vaginal yeast infection*

DO NOT continue taking the beverage without consulting with a health practitioner. If no qualified health professional is available to you, **DO NOT CONTINUE DRINKING THE**

KOMBUCHA BEVERAGE. There could be serious medical effects if you ignore this warning.

If you have not suffered any of the above problems, you may then choose to increase your daily consumption. Always drink the beverage before meals. The quantity that is consumed **should NOT exceed 8 ounces per day** unless supervised.

This recommendation is for a reasonably healthy individual with no outstanding medical conditions. Because of the impact Kombucha beverage has on the liver, excess amounts can cause problems.

THE TEMPERATURE...

The yeasts are hardy and can survive cold temperatures. The bacteria are more sensitive to temperature fluctuations. If you have attempted making your own yogurt, you have had experience with this. Between 72 to 80 degrees F. is the optimal brewing temperature range. The need for a rather warm air temperature can make the brewing process much slower during cooler months.

Some find that placing the brewery in higher locations in the house, in lofts or upper shelves of linen closets, can help meet the brewery's temperature needs. DO NOT place the brewery in an oven, assuming that the pilot light will provide the proper amount of heat. Because the oven is a closed environment and the pilot light is burning oxygen, you will be depleting the culture of a vital resource—oxygen. If you would rather be a bit more scientific and more able to efficiently control the environment, there are tools available to accomplish the goal.

An aquarium heater bought at your local pet or tropical fish store is an inexpensive alternative. Place the "brewing container" in a plastic or rubberized pan which has been filled with enough water to completely submerge the heater. Place the heater in the pan and set the thermostat at 72-80 degrees F. The only drawback with this method is if the water level evaporates down and the heater is no longer submerged, the heater will burn out. Check the water level every few days to prevent this problem. If you need to remove the heater from the water, unplug it first to allow it to cool down _within the water_. This is an important step in preventing the heater from breaking. (see resources)

Also available for providing an adequately heated environment for your brewery are seed propagation mats generally obtained from specialty greenhouse catalogues (see resources) and in northern climates at some nurseries. The mats are completely sealed and waterproof and use very little electricity. The metal shelf allows the heated air to rise and circulate between the mat and the brewery, maintaining the even moderate temperature needed. If, however, you are unable to obtain either item or have no access to loftier heated space, just count on a longer brewing time, sometimes up to 2 weeks.

KOMBUCHA STORAGE...

If you desire to suspend your Kombucha production or wish to keep babies for future use, simply place them in one cup of the beverage and secure in a zip-lock bag or a non-metal covered container and place it in the refrigerator. The refrigerated environment will suspend the fermentation process and keep the culture preserved for a fairly long time. However, approximately every 3 days open the zip-lock bag or container and allow air to enter the bag or container to ensure a healthy culture.

THE NEED FOR AIR...

When brewing Kombucha, the greater the surface area that is exposed to air, the healthier the culture and product will be. The importance is to keep the fermentation an aerobic process as opposed to an unhealthy anaerobic one that will produce toxic byproducts.

THE FIRST WARNING...

Lack of oxygen will provide an environment for the anaerobic production of acetone. This toxic compound is grouped among molecules referred to as ketone bodies and will harm the kidneys. Acetone has an acrid sweet smell, the kind that wafts from nail salons and out of nail polish remover bottles.

The appearance of ketones has been encountered in commercially bottled Kombucha. Once the beverage has been bottled and capped, care must be taken to keep the beverage refrigerated at all times even during transportation, in order to avoid the production of ketones. Also, an expiration date of 3 to 4 days is advisable. After the bottle is purchased, store the beverage <u>uncapped.</u> The anaerobic fermentation process produces dihydroxy acetone phosphate. The capping of the product shuts off the oxygen and acetone is trapped. If your beverage smells at all acrid or "chemical," *<u>DO NOT DRINK IT!!!</u>*

NO NEED FOR LIGHT???

Kombucha does not need light. In fact, if exposed to the ultraviolet rays given off by the sun, the culture will die. A dim or unlit area is the best place for a brewing jar.

HAVING A BABY...

The growth or "baby" appears on the surface of the medium. First appearing as blobs of dull, transparent jelly-like substance, the new growth will spread across the entire surface of the liquid. Under optimal conditions, this process usually requires less than a week. Again, this is all dependent on the air temperature and the diameter of the surface area. After it has covered the area, it will start thickening to produce a "healthy baby" approximately 2 to 5 mm thick (1/16 to 1/8 inch thick) and resemble a soggy tortilla. If this process is left to itself, over time, the babies will fuse into layers called lamella and form an opaque beige to manila-colored culture. When the cultures are left to thicken unchecked, they form less viable cultures which do not produce a quality product. I have found the optimum 3 to 5 mm culture the most reliable and healthy. Bacteria native to the Kombucha culture tend to degrade as a culture grows older or is grown under adverse conditions. They tend to mutate into "Y" forms or oversized versions of themselves. Because of this phenomenon, it is recommended to replace your "mother" culture with a "baby" every 5 to 6 batches.

It is quite easy to separate the baby from its mother culture. Using *clean* hands, it can be gently pulled apart. If it has adhered, fusing with the parent, a sharp knife can be used to sever them.

SECOND WARNING!!!

Hear the sirens? Check the rear view mirror. See the flashing red lights? There seems to be a problem. Better pull over. Oh, I see. It's a bright neon yellowish mold growing on the culture. See it there? Oh look, there's another culture with a black mold on it. Know what those are? It is *Aspergillus fumigatus*, and perhaps *Aspergillus niger*. These can be **extremely** dangerous and can even be deadly! The molds are normally found in soil, manure and composting vegetable matter and float freely in the air. To be safe, you should not grow the culture in a room with potted plants, or near or downwind of compost piles. That screened window does not protect your brewery from these microscopic invaders, my friend, because your garden compost might be too near the house. Be very, very aware of this and if you ever see any shade of yellow, black , or any suspect growths on the culture, dispose of the beverage and the culture. Obtain a new healthy culture and begin again.

ABOUT ACID AND METALS...

Acid coming into contact with metal for more than a very brief period of time can cause, by chemical reaction, the leaching of mineral salts from the metal.

The most eloquent example of this phenomenon was studied by nutritional anthropologists in the 1960's after it was discovered that certain tribal groups in Africa had developed iron-deficient anemia *en masse*. Research revealed the cookware the tribesmen were using had suddenly shifted from traditional iron cookware to aluminum and steel. Further investigation proved that the groups had received iron, leached from their pots while cooking high acid foods like tomatoes, providing enough iron necessary to supplement their otherwise iron-poor diets. This example is a positive aspect of metal leaching. There are, however, very negative effects achieved by the same process.

Eating of acid foods cooked in aluminum leads to a form

of aluminum poisoning known as heavy metal toxicity. Heavy metals are inappropriate metals of a similar electrical charge but of heavier molecular weight than the naturally occuring metaloenzymes found at nerve endings and tissue sites. The result is impaired function of those nerves and tissues. Our modern world has dramatically increased the incidence of heavy metal toxicity, of which aluminum poisoning is one of the most common. Its possible implication in human disease and pathology is insidious at best. It must be emphasized that bringing any metal into contact with the acid liquid of Kombucha will cause the metal to be leached into the beverage and must be avoided.

A PROBLEM THAT'S A REAL "PEST"...

Fruit flies, vinegar flies—call them what you will—these pests are most prolific in the summer appearing from out of thin air to buzz the brewery. If they manage to get inside the brewery jar these pesky aviators will lay eggs upon the culture which hatch out into "worms " (maggots) and will make the culture look rather unappetizing. Covering the brewery with a cloth thick enough to keep out these flying annoyances and securing that cloth with a rubberband, or whatever, is essential to protecting the culture.

Ants can be kept out of the breweries by surrounding the brewery container in water. If you choose to use an aquarium heater in water that problem is solved.

WHAT ABOUT THE ANTIBIOTIC EFFECT???

This area is problematic. What type of antibiotic effect would be my question. Most individuals should not ingest any antibiotic, either pharmaceutical or natural, on a daily basis if it is in a penicillin-like compound. Yes, the taking of antibiotics for acne is a dangerous route of treatment, except in very rare cases, because it compromises the body for a short-term gain of blemish-free skin. Resistant strains of bacteria can appear from the casual, indiscriminate or overuse of powerful antibiotics. Also, iatrogenic disease conditions can arise, namely candidiasis, and the destruction of the gut flora without replacement causes degenerative digestive patterns. Once these patterns are established, the path can often lead to arthritic-like conditions and general malaise.

Some Kombucha books now on the market credit the beverage with containing a natural antibiotic called usnic acid. Usnic acid is a compound extracted from the usnea lichen. Lichen are somewhat different creatures consisting of algae and fungi and requiring light to live, different than the yeasts and bacterias found in the Kombucha culture. This combination of factors makes the Kombucha a very unlikely source of usnic acid.

Kombucha's antibiotic effect may come from its acid pH rather than from an active ingredient. Because of the acid nature of the liquid, it will inhibit certain bacterial activity. This type of antibiotic effect is not like penicillin as it is not *radically* altering the body's environment in its anti-bacterial action.

GETTING OVER THE SUGAR ISSUE...

Eeee Gads! White sugar!!! This is the bane of the health conscious and the dietetically evolved. The refined "carbo" menace has a surly, if not downright diabolical, reputation among health devotees. Surely there must be a better substitute?

Well no, not really. In this case, its reputation is a moot point. Once again, understanding the essentials of basic biochemistry provides the reason.

Cane or beet sugar is refined to make sucrose. Sucrose is a two-unit sugar or disaccharide, containing 1 molecule of glucose and 1 molecule of fructose. Glucose and fructose are monosaccharides. A monosaccharide is a 6-carbon sugar. The basic monosaccharides are: fructose, galactose and glucose. Of these, only glucose is used extensively by the animal and plant kingdom as the essential source of all cellular energy production. If the body did not convert more complex substances into this basic unit of glucose there would be no life as we know it. Unlike the human body, the simple structure of the yeast and bacteria cells have no need for the production of a carrier molecule for the glucose to begin its work. In our bodies, insulin is that carrier, piggybacking glucose through the cellular gates into the inner sanctum of the cell. It is radical fluctuations in insulin production, induced by sugar, that gives the sweetener its negative health impact. And although it is true that in humans, fructose requires less insulin to be carried in (approximately one third of the amount glucose needs), in the Kombucha beverage it is inconsequential, as both the fructose and the glucose are being transformed into CO_2, acetic acid and trace amounts of ethanol (0.5%, about the same as most commercially produced fruit juices) before an individual consumes it. If made properly this beverage has very little, if any, negative impact on blood sugar regulation. *However, this beverage is not recommended for alcoholics, recovering alcoholics, individuals suffering from cirrhotic liver conditions or severe diabetic conditions without the supervision of a health professional.*

SO, HOW ABOUT FRUCTOSE?

If you will remember back at our factory, Gladys (glucose) went right to work while Frank (fructose) was placed in a training program to be made over. Glucose can be readily taken into the cell but fructose has to go through a rearrangement of its structure before it can enter into the fermentation process or the energy cycle.

Because fructose molecules are slower to enter into the process, they provide a raw energy source for an extended period of time. It has been found that the actual glucose content of the beverage 3 to 5 days into the fermentation process rises as the fructose is being rearranged. This provides an extended regulated supply of glucose, or in the factory setting creates a steady supply of capable employees. But when no glucose has been supplied as it is in the sucrose molecules, the yeast portion of the process is delayed and the different beneficial acid byproducts are produced at reduced ratios.

The production of gluconic and glucoronic acids are diminished as they are dependent on glucose availability. Because fructose conversion slows the production of these beneficial acids, its use alone reduces the concentration of these acids in the beverage. With a sole source of fructose, the fermentation process is more likely to produce ethanol than these useful acids. Therefore, fructose is a less-desirable substitute for the sucrose and its exclusive use is not advised. If you are committed to the use of fructose, it is essential to provide a portion of the previous culture's reserve liquid to stimulate the acid and "jump start" the fructose conversion. However, if the temperature is too cold, the process has to be closely monitored, because the combination of these two factors (fructose and sub-optimal temperature) slows down the

Kombucha culture's processes.

Brown sugar, whole cane sugar, molasses, maple syrup and honey all contain impurities which can cause inappropriate fermentation processes or alter their functions so much that you are in fact creating something other than the "real " product. Experiments have been conducted with these ingredients and much inferior products have resulted.

Brown sugar is made from refined white sugar and gets its color from the addition of molasses or caramel. Its content is around 96% sucrose and 4% molasses. The only possible benefit from the use of brown sugar is very minute amounts of sulfur that would be provided by the molasses, but it increases the risk of adulteration with impurities.

Raw sugar is produced as a mid-production product of the white sugar refinement process. Because it is manufactured in third world countries, where concern for cleanliness is perhaps not as high a priority, this form of sugar offers the greatest possibility for contaminants.

Whole cane sugar and molasses are unrefined and impart very different characteristics to the Kombucha beverage when compared to the refined white sugar product. The high-sulfur content of both products will stimulate the torula portion of the yeast colony producing a rather thick yeast sediment on the bottom of the container. The color of the brew will be dark and cloudy with a distinctive malty flavor. The natural mineral content of the unrefined cane sugar imparts a good range of minerals to the brew, but it seems to inhibit the ability of the acetic bacteria to convert the medium to the proper pH in the normal period of time (6 to 8 days).

The use of honey is also fraught with problems. The proportion of specific sugars (fructose, glucose and a number of more complicated sugars called ogliosaccharides) vary according to the source of the pollen the bees have used to make the honey. Consequently the amount of honey that one would use to make the Kombucha beverage would fluctuate according to its glucose content. Because of this complication, it is difficult to use honey as the sweetener.

WHAT ABOUT HERB TEA?

How about using herb teas? Well, if the herbs contain volatile oils, which generally they do, active ingredients in the oils are released into the solution and the Kombucha beverage is greatly altered. Perhaps the most exciting phase of Kombucha use is yet before us with the experimentation of adding herbal teas, both native and Chinese, to brew the culture. Qualified practitioners conversant in the healing properties of various herbs could create remarkable medicines. However, only individuals with *extensive* knowledge of herbs with a biochemical background should embark on this endeavor as possible permutations are endless and the results could as easily create poisons as they could potions.

And now it is time to move to the next bug-a-boo about the Kombucha medium—caffeine.

THE BLACK TEA ISSUE...

In the beginning the word was with tea and all tea was green. Mass production of black tea required equipment that did not become available until the Industrial Age.

Green tea and black tea both come from the same plant (Camellia sinensis). Treatment of the leaves after harvest is what makes the difference in the qualities of each type of tea.

Green tea is treated simply. Freshly picked leaves are

exposed to heat, either steam heat, as they are in Japan, or dry heat which is the Chinese method. This heating process destroys chemicals native to the plant which would otherwise cause the leaves to begin fermenting. This process, however, leaves the tea with a bitter taste (high tannic acid). Immediately after this, the leaves are rolled and dried. Grading of Chinese green tea is determined by passing the finished tea through sieves. These grades include hyson, gunpowder, imperial and pea leaf. *Bancha*, Japanese green tea, is the most commonly found type of green tea available in the U.S. Very small amounts of caffeine are found in green teas.

Black tea requires 4 steps of production. First the leaves are spread very thinly on trays and allowed to wither naturally for 12 to 24 hours. Next, using special machines that essentially chew up the leaves, the broken cell structures of the tea leaves release juices and enzymes which spread over the surface of the leaves and initiate the fermentation process. The juices that dry on the leaves contribute to the flavor of the tea. This fermentation step initiates the chemical process which stimulates the alkaloid, caffeine, (a bitter nitrogen-containing alkaline substance) to emerge. The leaves are then spread on tables and allowed to absorb oxygen for 2 to 4 hours. Next, the leaves are moved to special rooms in which a 95% humidity is maintained and the temperature is regulated. It is here that the tea leaves are fermented.

The best-flavored Indian teas develop when the fermentation temperature is maintained between 75 to 84 degrees F. In China, fermentation is carried out at higher temperatures by spreading the rolled leaves in the sun or by forming them into balls that are kept overnight in a warm moist place. During this step, the leaves change from dark green to a bright coppery color and a musty-leafy odor is converted into a fruity one.

Finally the leaves are dried, usually at high temperatures to seal in the flavors. A second drying at a lower temperature removes the last traces of moisture to prevent spoilage.

These procedures produce pekoe and souchong teas from which all other black teas have evolved, including Earl Grey and English Breakfast. The process of fermentation and drying raises the caffeine content of black tea in comparison to green tea.

CAFFEINE...

"Energy! I need energy." The cry echoes across the country. It is rising to "the best part of waking up," the booming enterprise of cappuccino bars, the office coffee pot, reaching for the green plastic bottle with Extra Strength printed across its label, motoring up to a drive-through for a Pepsi, purchasing a Coke conveniently supplied in the school cafeteria or a dispensing machine in the hallway or, if you're a health-conscious individual, purchasing energy lifts in bottles or convenience packets from a health food store. Coffee, tea, cola, Excedrin and "energy pills" (generally loaded with guarana and kola nut) are the most common sources of caffeine.

This everyman's drug gives a lift by stimulating the central nervous system, prompting stored glycogen release from the liver which provides more glucose to the brain, heart and nerves for energy production. It is a potent diuretic which often also contains tannic acid and can cause the kidneys some trouble. Repeated overuse of this drug will exhaust the system, pure and simple. It is socially and psychologically habit-forming. Caffeine, just as with all things in life, should be used in moderation and without dependency.

Caffeine and caffeine's cousins, theophylline and theobromine in trace amounts, are all found in black tea and belong to a class of chemicals called purines. These purines break down first to xanthine. It is in the xanthine phase of its breakdown that caffeine stimulates all the reactions associated with its consumption, i.e. rapid heartbeat, diuresis and stimulation. In the final breakdown, xanthine is reduced into uric acid. Uric acid has two noteworthy attributes. It is the cause of gout, a swollen, painful arthritic condition that usually occurs in the joint of the big toe of its sufferers. Also the relative high amounts of uric acid in the human body have been attributed to our living longer than other mammals.

Theobromine and theophylline are both found in small amounts in black teas and are related to caffeine because of their xanthine structure. Theobromine is a diuretic, muscle relaxant, heart muscle stimulant and vasodilator. Theophylline has a stronger stimulatory effect than caffeine on the heart and the respiratory system. Along with aminophylline (another caffeine derivative), theophylline constitutes the active ingredients of the thigh cremes that have recently stormed the cosmetics market.

Against a backdrop of known detrimental effects of caffeine, rises the issue of caffeine in the Kombucha beverage. Again biochemical principles can explain its utility.

If an alkaloid is placed in an acid environment, it is transformed into other substances referred to as salts. These salts can have wide-ranging effects. Possibly the production of these caffeine-derived salts could provide insight into Kombucha's function as a potent liver detoxifier. Recent reports in cancer research suggest that both green tea and black tea can change the metabolic pattern of the liver and inactivate environmental toxins and carcinogens. Specifically, it was found that black tea, with its elevated caffeine levels (with the possible inclusion of its cousins, theobromine and theophylline), offered more protection than green tea against skin cancer in a controlled study conducted on mice. The specific mechanism by which this phenomenon occurred is not clearly understood but it is known that all teas, including decaffeinated versions, display antioxidant properties.

A tannin referred to as EGCG or epigallocatechin gallate was identified a number of years ago and was designated as the anticarcinogenic element in green tea causing its consumption to rise dramatically in this country. EGCG is a potent free radical scavenger which guards against tumors by attacking those insidious oxidizing radicals of free singlet oxygen that have been linked to degenerative conditions ranging from heart disease to cancer and aging. (It is free radical activity that turns the leaves of deciduous plants into their brilliant colors and causes them to drop off their branches in autumn.) However, Dr. Zhi Y. Zong of Rutgers University feels that too much emphasis has been placed on green tea and its high EGCG content. The new findings show black tea and its higher caffeine content was the more effective treatment. Because black tea is chemically more complicated than its green tea cousin, additional studies are being conducted to determine its potential contribution to cancer therapies. This is not a license to dramatically increase your caffeine consumption. All the evidence has not been considered. Please, moderation in all things is best.

THE MOST RECENT
YEAST DISCOVERY...

Saccharomyces cerevisiae yeast is the darling of DNA

research scientists. Because of its elementary cellular structure, it is a relative breeze to dissect the double helix and experiment with placements and rearrangements. Efforts to manipulate the yeast cell's DNA is fundamental to viral research. Many viruses including the cold virus are "so simple" in their structure that they only involve RNA structures which are surrounded by protein shells. In 1994, Dr. Asim Dasgupta, a professor of microbiology and immunology at the University of California in Los Angeles, made the discovery of a hidden anti-viral molecule within the saccharomyces yeast. Made of RNA, the component which inhibits the replication of the invading picorna rhino virus (responsible for the common cold) it is also proving to be effective against polio, hepatitis A and the Coxsackie virus (one of many viruses involved in upper respiratory infections).

No, this does not mean that nutritional yeast gravy is a cure for the common cold or that sniffing baker's yeast will quell that pesky runny nose and congestion. The process for extracting the working unit is very "high tech," and researchers are as yet uncertain whether its ability to stop replication of these viruses might have detrimental effects on other cells as well. The discovery is significant in that it startled the research community by its simplicity. Dr. Dasgupta's finding has important implications in one of the perennial quagmires of medical research, the treatment of viral infections.

A PANACEA???

Call me a spoilsport, a cynic, a non-believer, a "thou of little faith." I readily concede that I do not believe in panaceas. Healing is multidimensional. No *single* balm, potion, or therapy takes the place of taking total personal responsibility for how one lives one's life. How much respect you show yourself, your body and your psyche is reflected in your state of being. No one thing can instantly and completely remedy years of abuse.

IN SUMMARY..

Kombucha's most dramatic benefit is as a liver detoxifier. By understanding traditional Chinese medicinal principles regarding the liver's connection to disease and conditions as wide ranging as hay fever and arthritis, one can perhaps grapple with some of the far- flung healing claims made about the beverage. Affecting the liver by "smoothing" out its function, the beverage should enhance this vital organ's ability to perform its innate tasks including:

- *stimulating metabolic rate (more energy generated by the ability of the liver to use thyroxin more efficiently, increase production of red blood cells and improve the oxygen-carrying capacity of the blood)*

- *ridding the body of toxins (which would diminish lethargy, depression, irritability) stimulating diuresis (reduction of excess fluids)*

- *fortifying ligaments and tendons (increased flexibility of the joints and reduction of arthritic pain through the increased ability to articulate the joints)*

- *reducing allergy-like responses involving phlegm and mucous production (improvement of the liver's ability to scavenge incomplete protein products [IPP's] which then would not enter the blood stream).*

- *improving the immune response (by stimulating the liver's release of bioactive vitamin A, [retinol] needed for the proper structure of epithelial cells of the thymus and mucosa, two sites associated with the production of immune factors).*

- *The effect of lactobacilli in creating a proper bowel environment could contribute significantly to all these factors, because when the body is digesting, assimilating and eliminating more efficiently, more nutrients can enter and nourish the body.*

- *The presence of theophylline in the black tea medium suggests the ability to stimulate cellular respiration and cause the body to lose fat and perhaps cellulite. Coupled with this improved fat metabolism are suggestions that improved fatty-acid utilization on deficiently functioning sebaceous glands could slow or halt hair loss. The B-vitamin content of the beverage provides elements that could help maintain the hair's youthful color.*

Each person's experience with Kombucha will be different. It is very unlikely that one will experience all of the possible benefits. How the beverage will work depends on one's over-all state of health, genetics and physiological idiosyncrasies.

If you have intuitive faculties, please listen to them. If you feel the urge not to drink Kombucha, if the body is somehow signaling to refrain, please listen. Your intuition is your most valuable counselor. Even water is deleterious if too much is consumed. Be aware. The bottom line is that Kombucha can help the vast majority of individuals. Almost everyone needs to do his or her liver a favor. This beverage could accomplish just that.

A SPECIAL WARNING TO AIDS PATIENTS...

I found that *all* AIDS patients I have consulted have severe candidiasis. Because of the immunosuppression which occurs from these virulent candida infections, I recommend that AIDS patients **DO NOT USE KOMBUCHA**. The yeasts inherent to the Kombucha culture and beverage **will aggravate** this very dangerous situation. At the very least I recommend that AIDS patients consult with competent practioners who are well acquainted with systemic candidiasis and immunosuppression to be closely monitored before experimenting with the Kombucha beverage.

A THEORY CONCERNING KOMBUCHA'S ENERGY-PRODUCING BENEFIT...

The next time you find yourself in a mixed-age crowd of individuals, perhaps at a party, or a busy weekend at a park or zoo, conduct a simple field study. Break the age groups down in terms of: toddlers (walking to 4), children (4 to 12), adolescents (12 to16), young adults (16 to 35), adults (35 to 65) and seniors(65 +) and observe activity levels of the respective groups. Energy levels are generally determined by metabolic rates, but I propose a deeper theory. It could explain how a two year old can run even a young mother "ragged." It could answer how your adolescent son's legs have developed wheel attachments with which he cruises 15-mile circuits of the neighborhood and shopping center from dawn to dark, or, why a junior high school teacher will have a difficult time keeping control of a classroom while lecturing on the French Revolution. Generally, activity levels of children are significantly higher than the adults who must care for them. It definitely has to do with the liver and could have a great deal to do with the level of a specific enzyme utilized by the liver called alkaline phosphatase.

Alkaline phosphatase [alk phos] levels are interesting in how they fluctuate during one's lifetime. Let's begin with a pregnant woman's level. During the last trimester, her alk phos levels will rise to twice her normal levels, a welcomed relief after the energy drain of early pregnancy. When she begins labor, her levels will rise to 3 times her normal level. A midwife's reliable signal of impending labor (a double entendre) is the pregnant woman's unusual spurt of energy which usually results in a thorough cleaning of her house a few hours before her other labor commences. The mother's alk phos levels will return to her normal level within 2 to 3 weeks after delivery. However, in the first four weeks of life, the baby's alk phos level rises rapidly to 5 to 6 times normal adult value. It will descend slowly until puberty, when once again the enzyme level will rise significantly, then decrease to normal ranges after puberty. These demarcations of elevated alk phos levels coincide with growth and maturation phases. It is in the adult phase that the alk phos level is at its "normal " value, rising slightly in the elderly.

According to the *Merck Manual*, normal ranges of alk phos vary between 30 to 110 mu/ml. If one is able to increase within normal ranges one's alk phos levels, it could be a mechanism for increased active energy output. Alkaline phosphatase requires a high alkaline pH in order to function. This friendly environment can be achieved in a healthy detoxified liver when it has been rid of acid toxins. The Kombucha beverage assists detoxification with its glucoronic acid, its methionine and cysteine content and its acid pH. However, increased alk phos values above the 110 level are indicative of very serious disease conditions. Again it is a moderate rise within one's normal range (e.g. from 45 to 60 mu/ml) that provides the maximum benefit.

PLEASE TAKE NOTICE

If the raising of alk phos levels is the mechanism Kombucha utilizes, it would follow that the ***EXCESSIVE*** *USE OF KOMBUCHA COULD LEAD TO FATAL OBSTRUCTIVE LIVER DISEASES like those reported and being investigated in Kombucha users in South Africa, as well as diseases of the pancreas, lung and bone.*

Human beings are notorious for the "if a little is good, a lot must be better" syndrome. Stressing moderation is the only way to guard against this phenomenon. Please take responsibility for yourself and be judicious with the use of this product.

When a situation like the arrival of a "miracle cure" arises, the skeptics and powers that be are waiting for the fall (someone to be harmed) and tend not to be interested in investigation or research after that fact. All too often the product is then made illegal to protect the public. Trained, responsible, alternative practitioners exercise a moderate approach backed by many hours of observation pertaining to how to treat a problem or a patient. Please, every individual needs to exercise caution and responsibility so it can remain a benefit without criminalization or excessive governmental regulation.

MUSCLE TESTING...

"Muscle testing" is a simple technique employed by some professionals, including alternative health providers. Based on the belief that the body has an intelligence of its own, this system attempts to tap into that intelligence. Through this method, you can communicate with your body by asking it a simple question that requires only a "yes" or "no" answer and testing for its response.

Muscle Testing

HOW TO MUSCLE TEST FOR KOMBUCHA...

Facing the person who will be testing (tester), the subject will hold her left arm at shoulder height, extending it straight out to the side of the body, palm down. In the subject's other hand, which is drawn up close to the chest, is a cup which contains a small sample of the tea. (see illustration above) The tester will place her right hand upon the back of the wrist area of the left arm of the person being tested with the tester's left hand upon the shoulder of the subject. The subject will ask aloud, "Is this tea good for my body?" The tester will then apply a steady force upon the subject's extended arm. The subject is to resist with adequate effort. If the response is "yes," the subject's arm will remain strong and extended outward. This is taken to mean that the body would find the tea beneficial. However, if the answer is "no," the subject's arm will drop or weaken significantly and it will be able to be

73

pushed downward more easily by the tester, despite an adequate attempt at resistance by the subject. This response is considered to be the body's way of signaling that it would not be to its benefit to use the tea.

An interesting experiment can provide evidence of this form of testing and the body's intelligence in action. As the subject, extend your arm and ask yourself, "my name is (use your correct name)." The tester will be unable to push your arm down. Then, state "my name is (use a fictitious name)," the arm will not be able to resist the downward press applied to it. Remarkable, but try it. Important to this process is not to abuse the procedure. The test is not a test of one's strength. Resisting is a cursory resistance to the force that the tester is applying. The tester as well should not strive to use undue force to bring the subject's arm down. Muscle testing can be a valuable tool for those of you who want to try it. It's akin to using your extended arm as a dowsing rod, so to speak.

QUESTIONS COMMONLY ASKED CONCERNING KOMBUCHA...

Q: I store my Kombucha beverage at room temperature and it keeps growing babies on the top of the surface. Is that O.K?

A: *Despite the most efficient filtering system you could dream up there is little likelihood that you would be able to eliminate all the yeast and bacteria cells from the beverage. Saccharomyces yeasts are very prolific even when replicating from a single cell. It only takes one yeast cell to stimulate the culture growth. This is not a problem, just the nature of the brew. Refrigerated storage of the beverage will hinder further reproduction.*

Q: Can I eat the actual culture?

A: I would not do it nor do I recommend it. I have interviewed individuals who have. They did not find it nearly as appetizing as the beverage and did not repeat the event. If, however, you wish to embark on this culinary journey, I recommend an abalone pounder to tenderize it first.

Q: Sometimes my tea comes out sour. What does that mean?

*A: All Kombucha beverages are a little sour in their "vinegary" sense. If it is not a vinegar-like sour flavor or it is too sour, I would consider the possibility that some form of contamination has occurred and would not drink the "batch." If the same culture is used again and the same problem occurs, dispose of that culture, obtain another and begin again. The possibility exists that the culture's yeast and bacteria ratio could have been altered, producing excess acids and you may be brewing an all together different beverage. If your beverage ever displays any unusual characteristics i.e. "strange" smell, "strange" look or "strange" taste, **DO NOT drink it.** Try another brewing process. If the same problem arises, once again, follow the advice mentioned above, be exacting in following the directions and, hopefully, you will brew a successful batch.*

Q: Can I flush a culture into the septic system?

A: Sounds like a future X-files episode to me. Quick, call my agent. Beyond that, I would assume that the very "hot" environment of the septic tank would "burn" the culture. Kombucha requires carbohydrates (sugars) to grow. Carbohydrates are the most thoroughly extracted nutrient removed during the digestion process. The amount of available carbohydrates for the yeasts to grow would be negligible. However,

I would reason that the bacteria aspect of the culture could contribute to the breakdown processes that occur within the system. I do not see this as a problem.

Q: Can I put the Kombucha in my compost?

A: *I would not compost the culture. I have concerns pertaining to the subsequent increase of yeasts in the soil. The soil has specific levels of activity concerning its constiuents and the nutrient milieu of the soil community. Tampering with this rather delicate symbiosis is best left to organic soil management experts. I have been an organic gardener for over 25 years and, having been thoroughly impressed with the book* The Secret Garden *by David Bodanis (which describes the drama under our feet), I have arrived at this opinion.*

Q: I'm pregnant. May I drink the beverage?

A: *No! The Kombucha beverage appears to function primarily as a liver stimulant and detoxifier; the risk of putting toxic compounds into the blood and some of those toxins passing through the placental barrier are too great. The consequences could be serious. The same warning is extended to nursing mothers, as well.*

Q: How about Kombucha for my toddler and adolescent?

A: *No. My concerns lie with the theory of increased alkaline phosphatase levels and with their connection with developmental processes. I personally do not give it to children and I do not recommend it until someone has become an adult and has reached full maturity.*

Q: Do you adjust the quantity of Kombucha beverage to be taken according to the relative size and weight of the individual drinking it?

A: No. Weight and body size is not as much of a factor as the relative "metabolic health" of the individual. Metabolic health would be determined by the amount of toxicity the person exhibits.

Q: When is too much, too much?

A: If the body is overly toxic the discharge of the toxins into the bloodstream would cause uncomfortable symptoms, i.e. headaches, nausea, "groggy feeling" and bowel problems. I recommend starting at modest levels of consumption to offset any possible complications. Increasing quantities in 2 ounce increments until you have reached the recommended 8-ounce daily dose is an intelligent way to reduce the risk of adverse side effects. Increasing the daily consumed amount above this 8-ounce level could be dangerous. Using Kombucha over a longer period of time and at the recommended doses is a responsible method. Trying to "force" your liver by overdosing, exposes you to too many possible consequences that could be harder to remedy than the original toxicity.

Q: I have a friend who tried Kombucha and became nauseous. Now I'm afraid to try it.

A: Your friend could have been overly toxic, taken too large of an initial dose, or had candidiasis or an ulcer. Any of these conditions or situations could have caused the problem. This is why Kombucha is not for everyone. If you start out at the recommended level of 2 ounces and do not have any of these pre-existing conditions, there should not be any trouble. Muscle testing is an approach used by some to access the body's innate intelligence on the matter.

Q: How can Kombucha be a liver detoxifier if it contains alcohol?

A: As explained in the text, Kombucha contains no more alcohol than commercially bottled fruit juice or non-alcoholic beer. You might wonder how alcohol comes to occur in fruit juices. Yeasts are so prevalent in vegetables and fruits that the natural yeasts that occur in the bruised and damaged fruits used by the juice industry have started fermenting before the bottling process has begun.

Q: Speaking of alcohol, I have a friend who uses vodka as his starter. He says it's to provide an "alcohol sugar" to start the fermentation process. Is this O.K?

A: Increasing the sugar content—which you say is the reason your friend is using vodka—is not the reason for using a starter. The starter is to increase acidity, making the medium's environment that much "friendlier" to the culture. If you do not have the finished Kombucha product to add to your next batch, vinegar—white or apple cider—is an acceptable substitute.

Q: How many times can you reuse the Kombucha culture?

A: I recommend changing your culture every 5 to 7 batches to a younger version. This is based on research which revealed that yeast cells can undergo "mutations" the older they become.

Q: Why do my "babies" grow thinner in the winter?

A: Yeasts are pretty hardy characters and are not as dependent on a properly maintained temperature as the bacteria are. Consequently, replication of the culture is stemmed by a significant disruption in the ratio of the culture's constituents. Increasing the temperature in the brewery to within a 72 to 80 degree F. range should facilitate reproduction.

Q: What are those tentacles hanging from my culture?

A: Ah...Thanks. More material for the X-files episode I have in development. Actually, they are dead and dying yeast cells and pose no problem.

Q: What is the sediment on the bottom of my brewery jar?

A: Yeast. Actually, yeast carcasses. They are resplendent with nutrients and should be allowed to remain in the jar for a few brewings if you are immediately restarting a culture in the same jar. Do not save them if you are not starting a brew immediately. Rinse the jar every few batches and begin again.

Q: My Kombucha foamed when I poured it out. Is that OK?

A: Yes. To determine if you are a consummate Kombucha brewer, measure the "head" it produces when you pour it out. A really healthy batch is 3 inches or more when poured into a clean, well-rinsed pitcher. The foam is carbonation, which is carbon dioxide released from its carbonic acid state because of exposure to oxygen. These are the pleasant champagne-like bubbles inherent to the Kombucha beverage.

Q: When I first started brewing Kombucha I was given instructions that emphasized that I should only use glass or ceramic utensils to boil the water for the tea. I see that you are saying that it is OK to use stainless steel. Why?

*A: Bringing water and sugar to boil in stainless steel causes no leaching of undesirable metals. Bringing acid into contact with it could. The acid nature of the Kombucha beverage means that it has to be fermented in glass or ceramic contain-ers because of this leaching potential. **Do not allow metal to come into contact with the fermented tea.***

Q: I'm confused. You say that it is OK to cut the culture with a knife or scissors. Isn't this violating the metal rule?

A: No. This is because the culture is outside of its acid bath. Only a small part of the culture comes into contact with the metal for a very brief period of time and then the culture should be rinsed in cool water before placing it in solution.

Q: If I stop the fermentation process too soon, can I just put the culture in the tea and continue, or do I need to start over?

A: Simply replace the culture and continue on. The brewing process will continue from its existing pH to completion. If, however, you had tested the brew with a pH strip without removing the culture, the fermentation process could have been continued without interruption.

Q: How long does the Kombucha beverage last in the refrigerator?

A: Kombucha beverage stored in the refrigerator tends to go "flat" losing its carbonation. Except for this occurrence there is no problem storing the beverage in the refrigerator, covered but not capped, for a week. Capped, refrigerated Kombucha should not be stored for longer than 3 days because of the risk of acetone production. Acetone is a ketone and is very destructive to the kidneys. If you discover that your Kombucha has any unusual smell, in the case of acetones it will smell "sweetly acrid" like nail polish remover, **DO NOT drink it.**

Q: How long can the beverage sit on the counter?

A: The container of the fresh-brewed Kombucha should be covered to prevent contamination just as the original brewery is. The beverage will continue to ferment, increasing in acidity and carbonation. The rate of fermentation is dependent on the air temperature. It can be consumed for as long as it is palatable. Consequently, the length of time it may remain unrefrigerated is relative to your palate, usually for 5 to 6 days.

Q: If my "mother" culture did not rise to the top of the medium during brewing, is that all right?

A: *Kombucha "mothers" can be temperamental. My experience has taught me that it is disturbance or a temperature problem that causes them to remain on the bottom. The culture is buoyed by carbon dioxide trapped under it, raising it to the top of the medium. If the brewery has been disturbed, or the temperature surrounding it is not consistent, it tends to stay at the bottom. The critical issue concerning whether it is OK is determined by the beverage's pH. If the pH registers at the appropriate level (3.0), then the beverage is ready regardless of the suspension state of the "mother". However, the "baby" always grows on the top of the medium.*

Q: Can I make a double or triple "batch" by just increasing the amount of ingredients respectively? Do I have to use 2 or 3 "mothers" when I do this?

A: *In answer to the first part of your question the answer is yes. The second part of your question is no. It only requires one culture to bring even three times the volume of medium to the desired pH level. If you do use two "mothers," it could speed up your fermentation process by a day or so.*

Q: I read that brewing Kombucha in a glass jar does not allow enough oxygen to the culure but I see on the cover of your book that you are using a jar. What is correct?

A: *It is the circumference of the opening of the brewery that determines its ability to access oxygen. Any container with a 6-inch or larger surface area is sufficient. The jar in the* *photograph is a 3-gallon jar with a circumference greater than the 6 inches required. It is covered with an air-permeable cloth, not capped shut.*

Q: Why is a ceramic stoneware bowl OK to brew in and not earthenware?

A: *Earthenware has the possiblity of containing metals or lead in the glaze. Stoneware does not and therefore is safer to use.*

Q: I heard from my friend that Kombucha has really helped her PMS symptoms. Could this be true?

A: *Yes. Again this remedying of the symptoms of PMS is linked to Kombucha's ability as a liver detoxifier. In a liver that is "burdened," it cannot conjugate estrogen properly. When this happens estrogen is degraded into estrodiol, a noxious form of the hormone that is linked to PMS symptoms which include constipation. Because the patient is not able to discharge the estrodiol through the bowels, as would be normal, the estrodiol is taken up in the bloodstream and is recirculated causing irritability, headaches, bloating and swelling of the breasts. Kombucha alleviates PMS symptoms by improving the liver's detoxifying qualities as well as by acting as a bowel "sweetener" encouraging efficient bowel function. The Kombucha beverage's abilities as a liver detoxifier is employed in remedying menopause symptoms, as well.*

Q: As a matter of fact, my father told me that after his first dose of Kombucha, he had a "remarkable" bowel movement and has been regular ever since. What is a "bowel sweetener?"

A: *A bowel sweetener is a term that is used to insinuate that the bowel environment has been improved (sweetened) by a product. Pantothenic acid, lactobacillus and acidophilus* are all considered bowel sweeteners. Kombucha contains two of these products, pantothenic acid and lactobacilli. The more significant is the lactobaccilli which is provided by the bacterial element of the Kombucha culture.

Q: Can I recommend Kombucha to my friend who is undergoing chemotherapy?

A: I have recommended it to individuals undergoing chemotherapy because of its liver-detoxifying capability. Keeping the liver "loose" helps it to be able to deal with the assault of the chemicals on the body. **However, go very slowly. Start with 2 ounces and monitor the body closely. Consult with a trained health practitioner before beginning. If your friend feels at all "weird" after taking it, stop or reduce the dosage. This is another case where your friend could try muscle testing to help decide whether to begin using the beverage or not.**

Q: I have heard that I can use the culture topically on my skin. Is that true?

A: Thanks, the perfect final plot twist to my X-files episode. Actually, I have used both the culture and the beverage topically. The culture run through the food processor or blender and applied directly to the skin makes an invigorating facial mask. **However, its acid nature can burn some skin types.** *Buffering the culture mash with an egg white can raise its pH to a soothing mixture. If you choose to use the beverage on your skin* **as a toner, dilute it by one-quarter the volume with water to bring the pH up to a more skin-friendly 5 to 6 pH,** *thereby preventing any likelihood of burning your skin. I can reason that there could be some advantage to this topical use of the culture and the beverage based on its collagen-*

building acids. What is critical to whether any topically applied substance can actually plump up the skin and diminish wrinkles is relative to its ability to penetrate cell barriers. This is a factor cosmetic companies have been struggling with for decades and I cannot tell you if glucoronic acid can do that. As a toner, however, it has the potential to restore the natural acid mantle of the skin after cleansing.

Some individuals have reported that they also apply it to their cellulite. No one has reported back to me any astounding effect.

Another aspect of the topical use of Kombucha I found to be significant is applying the beverage as a mild topical antibiotic which has remedied conditions as diverse as mild acne to ingrown toenails. Apparantly, the acid pH and the various organic acid components of the beverage combine for this wondrous effect.

Q: Where can I get a Kombucha culture?

A: Networks of Kombucha "breeders and needers" have arisen due to Kombucha's growing popularity. Ask at a health food store or look in community alternative publications for potential sources. If, however, you are having trouble finding a culture, we have provided a name of a reliable source at the back of this book. The Kombucha Foundation has a database of "breeders and needers" so that they can connect with one another through listings. These cultures are so prolific, ask your friends; probably someone you know can help you find a culture.

⇜ PART 3 ⇝

The Bio-Chem Theater

FOR THOSE WITH A THIRST FOR KNOWLEDGE...

Fermentation is the principal function of the Kombucha Factory. Fermentation is by definition an anaerobic process. However, the bacteria and yeasts involved in the fermentation process of the Kombucha culture are considered "facultative anaerobes" which means that they grow in either the presence *or* absence of oxygen.

For simplicity, I have designated each division of the Kombucha Factory with a process. Because in microbiological terms yeast is considered a eukaryotic cell, it contains the "energy powerhouse organelle," the mitochondria, producing energy in the yeast division of the factory, which generates the Krebs cycle metabolites of citric, malic and oxalic acids plus yeast sediment (see illustration page 86). Bacteria are classified as prokaryotic cells which do not contain these organelles. The enzymes necessary to their function are found in the cell "soup" of their cytoplasms. In the bacterial division, the fermentation process takes place producing carbon dioxide and lactic acid, acetic acid or ethanol.

Discussions of these biochemical processes may sound

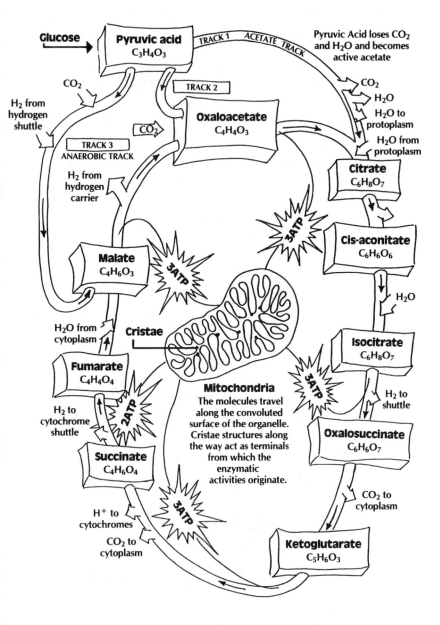

Glucose → **Pyruvic acid** $C_3H_4O_3$

TRACK 1 — ACETATE TRACK

Pyruvic Acid loses CO_2 and H_2O and becomes active acetate

CO_2

TRACK 2

H_2 from hydrogen shuttle

CO_2

TRACK 3
ANAEROBIC TRACK

H_2 from hydrogen carrier

Oxaloacetate $C_4H_4O_3$

CO_2
H_2O
H_2O to protoplasm
H_2O from protoplasm

Citrate $C_6H_8O_7$

3ATP

Cis-aconitate $C_6H_6O_6$

Malate $C_4H_6O_3$

3ATP

H_2O

H_2O from cytoplasm

Cristae

Isocitrate $C_6H_8O_7$

Fumarate $C_4H_4O_4$

3ATP

H_2 to shuttle

H_2 to cytochrome shuttle

2ATP

Mitochondria
The molecules travel along the convoluted surface of the organelle. Cristae structures along the way act as terminals from which the enzymatic activities originate.

Oxalosuccinate $C_6H_6O_7$

Succinate $C_4H_6O_4$

3ATP

H^+ to cytochromes

CO_2 to cytoplasm

CO_2 to cytoplasm

Ketoglutarate $C_5H_6O_3$

Krebs Cycle Main Chemical Terminals
En route to Aerobic Oxidation of Glucose.

like Greek to most individuals. Nevertheless, I will attempt a briefly outlined scientific explanation.

Living things obtain energy from carbohydrates. They extract energy from sugar (mainly glucose) by means of biological oxidation. Biological oxidation (glycolysis) is the chemical dissection (breaking apart) of carbohydrate molecules releasing energy for use. First, the glucose or fructose 6 phosphate molecule is reduced to pyruvic acid. When this acid comes into contact with enzymes, either in the mitochondria of the yeast cell or the cytoplasm of bacteria cells, the production of energy begins. The body produces an "energy carrier" called adenosine triphosphate (ATP) which interacts with these molecules producing heat, carbon dioxide and water. Fermentation by the bacteria of the Kombucha is, by definition, an enzymatic conversion of glucose to carbon dioxide and lactic acid, acetic acid or ethanol which results in the production of the same energy-storing molecule ATP. Fermentation in the bacteria division, however, is a very inefficient source of ATP yielding only 2 ATP molecules as opposed to at least 32 ATP molecules produced during the Krebs cycle.

With that said, why don't we explore this a little more creatively? Here's a "dramatic" explanation.

Welcome to...

➤→

THE BIOCHEM THEATER

A METABOLIC MYSTERY

—❖ THE SAGA OF ❖— FRANK AND GLADYS SUCROSE

CAST OF CHARACTERS:

Gladys Sucrose—(a glucose molecule)—the glucose part of the Sucrose Couple recently employed by the Kombucha Factory.

Frank Sucrose—(a fructose molecule)—the fructose part of the Sucrose Couple entered into a training program in order to begin working in the Kombucha Factory.

Ase N. Zyme—(an enzyme)—the magician. This conjuror will transform all those who come into his presence without himself being altered.

Kinase N. Zyme—(another enzyme)—a bus driver whose relationship to Ase plays a vital role in the mysterious events.

The Zoot Suit—(carbon dioxide)—a tap dancer whose greatest dream is to form the fabled Carbonic Acid Dance Team.

The Ballet Duo—(a water molecule)—the dynamic pair of OH^- and H^+, who are the elusive co-members of the Carbonic Acid Dance Team.

The Phosphate Brothers—(phosphate radicals combining to form adenosine triphosphate, ATP, and adenosine diphosphate, ADP; energy is produced when these substances are split by enzyme action)—powerfully built radicals whose dynamic dedication to energy is their driving force.

Princess Sophia Coenzyme of A—(an acetylizing agent SCoenzyme A)—the catalyzing regal presence which produces an enchanting environment for radical change.

Nic. O. Tinamide A.D.—(nicotinamide adenine dinucleotide, NAD, an electron carrier to the electron shuttle)—the rude and crude cigar-smoking waiter who gathers distressed birds for the "shuttle."

Flav N. A.D.—(flavin adenine dinucleotide, FAD, another electron carrier to the electron shuttle which contains the flavin element of the B vitamin riboflavin)—a yellow-skinned and equally rude and abrasive waiter who also gathers distressed birds for the "shuttle."

Cockatiels—(hydrogen ions, H^+ electrons)—distressed, squawking, wing-beating creatures made available to the electron shuttle by NAD and FAD.

Detective Will E. Findem—a private detective hired by the Kombucha Factory.

The scene opens in the office of a detective who is employed by the Kombucha Factory to investigate rumors circulating about the community concerning the disappearance of its employees. Despite these rumors, disappearances continue and yet more victims continue to flock to the factory for work. Recently, a breakthrough in the investigation has occurred, and the detective is bent over his typewriter furiously pecking away with his index fingers.

MEMO FROM DETECTIVE WILL E. FINDEM:

Mr. Frank Sucrose had reported to this office the disappearance of his wife, Gladys, and entered this letter as evidence of a crime of which we can find no additional evidence. Mr. Sucrose received the letter one week after his wife began work. The Sucrose couple had been recently employed by the

Kombucha Factory. Gladys was placed immediately into the yeast division (see illustration page 91) while Frank was placed in a training program for entering the bacteria division. We have had no luck in finding the missing person (either as Gladys Glucose or as Miss Pyruvic) based on the evidence obtained from the letter. However, we are concerned as certain curious facts discussed in this letter are verifiable by the local citizenry.

THE LETTER...

My dearest Frank,

Entering the factory was easy. There were no guards at the gate to check for passes and no fences to detain us. I settled back into the compartment of the factory's shuttle to enjoy the scenery of this mysterious place. We traveled past Golgi and past many endoplasmic reticulums. The kind lady next to me had a guide book from which she could identify the curious features and creatures we saw. Certain curious flat-bodied herds grazed unrestrained throughout the site. Shortly, we rounded a rather large group of the grazers and sped past the imposing skyline of the bustling metropolis, Nucleus.

We arrived at our destination, the Hotel Mitochondria. I looked forward to finally resting and maybe taking in the sights. But immediately, as I entered the lobby of the hotel, I was swarmed upon by urchins (enzymes) who giggled and snatched, pushed and shoved, and, before I could gather my wits, the urchins were gone. I looked down at my body and was shocked to discover that I had been reduced, reduced to half my original size. Oh Frank, it was horrible. I admit my thighs needed some modifying, but I did not need to lose half of myself.

Exhausted after this encounter, I took my weak and trembling body, and my indignation, to the concierge's desk. There I was surprised but, nonetheless, pleased to discover two other ladies

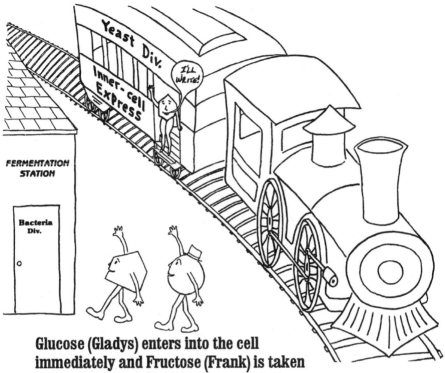

Glucose (Gladys) enters into the cell immediately and Fructose (Frank) is taken by enzyme to be altered

in line filing similar complaints. Together, with quivering voices, (remember, we all had been dramatically reduced to half our former constitutions by the assaults) we expressed our outrage. Dear, the concierge had the look of a weasel about him. Bowing and ingratiated himself in such an undignified manner, he reassured my fellow petitioners and me that we would be justly compensated by a complimentary tour of a nearby mountain range that he would arrange for us as the management's way of expressing regret for our outrageous treatment. I really did not like his manner and my instincts told me not to trust him, mostly because of one thing he kept doing which left me uneasy and concerned. He insisted on referring to all of us, individually, by the same name, Miss Pyruvic (a 3-carbon molecule). We all tried to correct him. I was surprised to discover both of my companions shared my first name, Gladys (glucose), but it was as if our objections were falling upon deaf ears.

My companions reluctantly agreed to the tour. As I did not want to be a naysayer and disappoint my new-found friends with a sour attitude, I too consented to the excursion and accepted the ticket for this fateful journey.

I was surprised that the tickets indicated we were scheduled to board the train immediately. The concierge whisked us out of his office and herded my companions and me through a door which opened onto a platform. As I stood upon the station platform, I noticed that the name emblazoned on the side of the shuttle was THE ENERGY EXPRESS. We boarded the train and settled into its one and only compartment. It wasn't a moment later that a large matronly woman addressed each of us as Miss Pyruvic, again, much to our distress. She identified herself as our tour guide.

She informed us that we were to be split up temporarily and ride separate shuttles on separate tracks. I expressed my concern, but she assured me that this was necessary and only temporary. I should have known something mysterious was afoot as one shuttle embarked to the west on track #3 designated the AN-AEROBIC TRACK, my other companion left on track #2 to the south, while I on #1, called the ACETATE TRACK, headed east. Each of these shuttles had only one compartment.

The following is an account of who came into my fancy large traveling compartment (uninvited, my love) and the process of transformation I underwent as a result. It was at first frightening and then became exhilarating for me to watch parts of myself fall away and, gradually, I became someone, no, something else.

As my shuttle pulled away from the platform, a rather short and pudgy young man inserted a tag in a slot on the compartment door and he entered. As the door closed behind him, I looked at the tag he had placed in the slot and it was labeled "Miss Pyruvic." Darling, I decided it was futile to protest. Besides, with the radical alteration of my appearance, I began to feel that a name change might be in order.

The young man presented his card of introduction identifying him as "Ase N. Zyme, a magician." He informed me that he would like to entertain me by taking an article of mine and transform it into something quite unique. The idea intrigued me, so I offered him one of my carbons. With a wave of his hand, there appeared a pair of ballet toe dancers. They introduced themselves as members of the Water Ballet Troupe (OH^- and H^+, H_2O, water.) Then the pair opened the door and leapt out.

The toe dancers had suddenly departed and now there stood a solitary tap dancer dressed in a canary-yellow zoot suit. He began tapping, his toes clicking in a staccato rhythm.

Spinning to face me he demanded, "Did you see where they went? I want so much to dance with them. I know I have the talent."

All the while the magician sat by with an amused expression. I stared at the tap dancer blankly, taken aback by these sudden appearances and disappearances occurring in my compartment. Seeing my confusion, the tap dancer smiled and asked my permission to seat himself upon my bench. He introduced himself as Carbon Dioxide. He took up my hand and held it in his own. His soothing presence calmed my troubled mind and, after a moment, I told him of the dancing pair's exit out the door.

He jumped to his feet shouting, "Wait, we can form a dance company! I even have a name for it. What do you think of the 'Carbonic Acid Dance Team'?" With that he leapt out through the door.

Immediately, two figures shadowed the compartment door. The magician rose and opened it. The two shadows proved to be identical twins. (Darling, I wish to let you know that I have always found you most attractive and pray my description of the two will not cause you too much distress, but certain qualities the pair possessed were so compelling that the only way I have to express it is in rather base terms. My dear, these were the two most handsome, virile charmers I have ever encountered. Their personal magnetism caused me to swoon. I believe the Latins have a term, *muy macho*, and darling they were the quintessence of that expression. I derived the greatest pleasure in that I did not have just one of these dynamic individuals in my compartment, but two, and identical in every way.)

They smiled broadly and with one magnificent tenor voice, they introduced themselves as "Phosphates" (DPN—diphosphopyridine nucleotide) and settled onto my bench. My face flushed at my thoughts as I struggled to pull up every reserve of my upbringing to avoid my lusty feelings, but, I am embarrassed to say, my demure breeding was forgotten as I squeezed myself as close as possible to the two studs.

Just as I had snuggled up, a knock was heard at the compartment door. I looked up to see a very tall and elegantly dressed middle-aged molecule of considerable beauty. Reluctant to share my bounty with anyone, I tried to ignore her. But my twin gods rose and with deep flourishing bows, they showed the lady in. Imagine my surprise when, as the lady glided into the compartment, the fellow in the canary zoot suit (carbon dioxide) slid in right behind her.

With the grace of a highborn, the lady surveyed the spacious compartment. My two darlings instantly scooted apart and she seated herself. Extending her hand, the lady introduced herself as "Princess Sophie Coenzyme of A" (SCoenzyme A). I kissed her hand as my two Adonises nodded their acknowledgement. The zoot suit tap dancer moved about the cabin mumbling and clicking his toes all the while. The Princess lifted her head and, by simply fixing her gaze upon the restless tap dancer, he slowed, stopped, turned and, for the first time, acknowledged his fellow travelers. In a moment he turned his gaze upon me.

Reminded of our previous encounter, he smiled sheepishly and asked to join us on the bench. We squeezed apart to make room as he wedged himself in beside us. The princess sighed and whispered, "Isn't this wonderful?" And it was.

After a moment I noticed that Ase had donned white gloves, turned his suit coat round to reveal a waiter's coat and with a bow and a flourish produced a frosty bottle of champagne. Suspended in midair, the bottle turned and poured its liquid into long- stemmed glasses that would pop into the magician's hand. My companions and I reached out to take the champagne glasses suspended magically before us and along with the words "In your honor, my Lady," we toasted the princess and took a drink.

"Psst... Ugh," I spat after taking a sip of the liquid. It was sadly overripe. It was vinegar, Frank. All the others drank it down with gusto. Embarassed, I too finished my glass so as not to offend the great lady.

Shortly thereafter, a porter passing our door looked in, paused and, reaching into his breast pocket, he pulled out a tag and displayed it through the window of the cabin. It read "Miss Acetyl Coenzyme A." The princess smiled graciously, reached out, took my hand and nodded to the porter. He inserted the tag into the slot. We all relaxed, basking in the warmth of our new-found camaraderie.

Ase, the magician, rose from his seat. With a twinkle in his eye and a smile upon his lips, he reached behind my ear and placed something there. Before I could raise my voice to query the chunky illusionist, the ballet duo (water) leapt through the door into my compartment.

The princess, acknowledging that the compartment was becoming overcrowded, bid us farewell and exited. Her depar-ture seemed to agitate all the remaining passengers and we began to jostle about. The magician reached out and pulled from thin air a tag labeled "Ms. Citrate," and once again placed it into the door slot. It was then I noticed that I had regained something of my previous bulk, but somehow it was distributed differently.

Now darling, events began to happen at a breakneck speed as the train began to descend the mountainside. The magician waved his hand as the ballet duo (water) again leapt out the door. This made Mr. Zoot Suit (carbon dioxide) very agitated and he began to prowl the compartment again. Click, click, Click, click. The magician pulled yet another tag from out of his sleeve and inserted it into the slot. This one read "Ms. Cis-aconitate." Now I felt balanced, every thing equal and in place. And what do you know? The toe dancers entered the room again. The pudgy little sorcerer reached into his vest pocket and presented another tag; this one read "Ms. Iso-citrate." Just as he placed this name in the slot on the door, the conductor, I assume, began shouting, "Tickets. Tickets, please." My two Fabio clones rose and, in one motion, flung up the compartment's outside window. To my shock and horror, they threw themselves from the train while shouting, "ATP, Adenosine Triphosphate!!!!" (surely some political slogan). Three great explosions with flashes of heat knocked me back in my seat. I began to agonize at the thought of my loss, but my anguish was cut short as three familiar shadows appeared at my door and filled me with delight. This time triplets of those magnificent buffed bodies entered my compartment. In one sonorous voice they began, "Let us introduce..."

I raised my hand interrupting their introduction. "Let me guess. The Phosphate Brothers?"

"Why yes," they replied in one voice, "of the family Triphosphopyridine nucleotide."

"What joy!" I shouted, just as the door flew open and a portly cigar-smoking waiter pushing a meal cart bullied his way in. His name tag read "Nic. O. Tinamide A.D." (nicotinamide adenine dinucleotide).

He pulled his smoldering stogie from between his fleshy lips.

"Well, well, well. This is a cozy group. Are the boys ready?" he inquired of the magician.

"Ready!" I jumped to my feet. "No! They just got here," I shouted as I threw my body across the path of my divine trio.

"Lady, lady, lady. What's the problem here? Hey, it's the boys for the shuttle I'm after." He turned to the magician. With a flourish that would make Siegfried and Roy proud, Ase N. Zyme reached in and pulled two squawking cockatiels (2 hydrogen electrons) from out of his sleeve for the electron shuttle.

"Ah. There they are," the cigar-chomping waiter sighed. "Oh, my sweet, sweet boys, Poppa's here." He lifted the lid of the covered silver dish on his cart and placed them under it. Opening the door, he pushed his cart into the aisle.

"What a hysterical broad," he muttered as he left.

"Broad!! Broad!!" I fumed. This time my three stud-muffins restrained me as I offered to fry ole' Nic's huevos.

Nic slapped a new tag in the name slot and bolted down the aisle. Now it read "Oxalosuccinate." "Do you believe his rudeness? No Mrs., Ms. or Miss." Oh, I felt insulted. I paced the compartment ranting to my companions.

Imagine my embarrassment when that fine lady, Princess Sophie Coenzyme of A, entered my door, bowed in by the Phosphate Brothers. Her presence was like a warm, soothing balm of peace and tranquillity. She placed her hands upon my shoulders and entreated me to join her on the compartment's bench as she seated herself.

Together we squeezed between Mr. Zoot Suit and the hunks as a sense of deja vu came over me. Shortly, the princess addressed me.

"Dear one," she began, "you do us proud. You have a right to dignity and respect for all your good work. I speak for all as I respectfully acknowledge your contribution." And with that she kissed *my* hand and smiled. The kind-faced porter again appeared and once again rummaged in his jacket's pocket and displayed a name tag through the window. It read "Madame Ketogluterate." The princess nodded her assent and the porter placed the tag in its slot.

I became aware again of the shuttle descending another mountainside at increasing speed as the princess begged her leave and left my compartment. I had only a moment before my ears were assaulted by a voice that caused me great anxiety.

"Tickets. Tickets, please," came the conductor's voice.

I spun around just in time to see the three bucks casting themselves out my window just as the twins had. Again the battle cry, "ATP, Adenosine Triphosphate!!!!" Three more explosions, three more flashes of heat and, once again, I crumbled desolate and despondent to the floor.

The wizard clucked his tongue at my despair. I felt exhausted. I have been transformed and rearranged so many times by now I hardly recognize myself. I'm sorry to say, my dear Frank, that what I was for you is difficult to find in me now.

As I sat there on my compartment floor contemplating my reality, I realized that Mr. Zoot Suit (carbon dioxide) was still present. I raised my tear-stained eyes as he moved to stand before me. He reached down and patted my cheek.

"I regret I cannot stay to console you. I have my career. I must go," and with that, he tap-danced his way out of the door.

No sooner had he exited, then the door opened and once again Mr. Zoot Suit's elusive quarry, the ballet pair, glided into my compartment. Close behind, two familiar shadows appeared. The pair entered my compartment and extended their hands. As is proper, they tried to introduce themselves, but before they could, I interjected, "No need. I am well aware you are Phosphates. I am pleased to make your acquaintance as I have meet your brothers."

The magician leaned out the doorway and placed another name tag in the overflowing slot. It read "Madame Succinate." He then pulled another squawking cockatiel (hydrogen electron) from his sleeve.

"Flav, Flav, I expected you earlier. Come. Here's another for

98

the shuttle!" he shouted down the hallway. A different waiter wheeled his cart to the door and he, too, placed a distressed bird under a silver coverplate. As he prepared to leave, he thrust his head into the room. I was shocked. I sat with my mouth agape. Darling, his skin was yellow. Yellow! Yellow as a daffodil, as the sun, as a lemon. I read his name tag, "Flav N. A.D." (flavin adenine dinucleotide).

"What's your problem lady? Ya' never seen a flavonoid before? You know, riboflavin? Geez, get a grip." His clipped high-pitched speech pattern was nearly as agitating to the ear as his skin color was to the eye.

"Say, are you the huevos dame?" His words jolted me out of my stupor.

"Dame! Dame! What is it with you A.D.'s? Your job description includes insulting? Well, I've just about lost my good humor. I've been assaulted, insulted, diminished, restored and rearranged. I'm exhausted and confused, and if I get one more smart-mouthed A.D. insulting me, I too will commit myself to whatever cause these fine units here are involved with and throw myself from this train. What a pretty picture that would be! Try to explain that to your bosses! I have documented each and every incident that I have endured. Heads will roll," I threatened.

"I'm sorry. We would not like that to happen," the magician interrupted pushing the brazen yellow-skinned waiter out and closing the door behind him. "We apologize and will endeavor to make the remaining portion of your trip an enjoyable one. Please be seated, relax, and let me take care of the situation." I noticed that the roly-poly wizard was sweating profusely as he tried to calm me. Concerned at the distress I was causing him, I returned to my seat between the twin Apollos and considered my situation.

Then once again came the faithful call, "Tickets. Tickets, please."

I was now resigned to my fate. Before they could disappear,

I grabbed both the twins.

"At least allow me this pleasure," I pleaded and pecked them both on the cheek. They exited through the window. Their cry, "ATP. Adenosine Triphosphate!!!!" Two blasts. Two flashes of heat. I sighed.

A rap at the door was heard. I recognized the voice of Nic. O. Tinamide A.D. and smelled the disgusting stink of his cigar. The magician, understanding my distress and endeavoring to allow me to retain a shred of my dignity, kept the waiter in the hallway as he once again produced two flapping, noisy birds from his sleeve for the shuttle. He replaced the name tag with a new one signifying my new status as "Madame Fumerate."

I was pleased to feel balanced and peaceful once again. Casually, I leaned back in my seat as slowly and gently the doorhandle turned and the Water Ballet couple reentered my compartment. They began a most sublime dance as three figures again shadowed the door. The phosphate triplets were back. I barely had the energy to respond to their magnetism. I begged them for no introductions.

I heard the whistle of another shuttle and rose to see my friend in the ANAEROBIC shuttle on a side track. I recognized a magician, a tapping Zoot Suit, a dancing ballet pair and the triplet shadows in her coach. The smile upon my friend's face told me she was finding pleasure in her company and had not yet been through the experiences that I had endured. I waved and she waved back as her train entered onto the track my own train was traveling. The ballet in my compartment continued as the magician again changed the name tags on my door. This one read "Madame Malate" (malic acid).

The conductor's voice once again rang out, "Tickets. Tickets, please." I merely waved to the trio as they flung themselves out the window. "ATP!!! Adenosine Triphosphate!!!" Three explosions. Three flashes of heat. Next time, I really must ask what the slogan means.

The magician pulled two more agitated birds out his sleeve. Nic was at the door and this time I was spared further conflict. As the magician attempted to close the door, Zoot Suit jumped into the compartment. "Ah ha! Here you are!" he shouted. But the look of excitement dropped from his face as there was no one but the magician and me remaining. The ballet dancers had just disappeared. I hadn't noticed their departure. So far, in my experience, only the phosphate brothers had used the window.

"Darn. Shoot. Fiddle-de-dee and fiddle-de-dum. I swear we would have made a good team. What's wrong with the 'Carbonic Acid Dance Team'?'" he lamented as he flopped onto the bench. I merely shrugged.

"I don't know what else to do. Would you mind terribly if I rest here with you?" he asked.

I told him that it was quite all right with me if he would do me a favor. The next time he left my compartment, would he mind taking a letter addressed to you, Frank, after I'd finished writing it, and get it to a post office?

Another shuttle whistle drew my attention. On another side track sat the shuttle my other friend had boarded. She appeared much more rested than I and had the anticipated companions within her compartment, a magician and a Zoot Suit. I waved as her train also entered the common track and we all moved on. It was then the magician tipped his hat and extended his hand to me. I reached out to shake it, but he surprised me as he put the back of my hand to his lips.

"You have endured and we all are grateful. There is a particular energy who is eternally grateful for your indulgence and sacrifice. It personally cannot thank you, so, as its representative, I wish to convey its greatest appreciation."

How curious, I thought...

"May I inquire what this energy is that I have inadvertently served?"

"Why, yes. It's Life! In this case, here within the Kombucha

Factory, you have ensured the budding of a baby, produced yeast sediment and contributed some of your acidic personas to a grateful patron. We are in your debt."

I nearly fainted. Imagine me, a servant of Life. I was so honored and proud.

And so, my darling, I am committed to my new service. I am poised on the track to repeat the adventure. All evidence of my past name changes disappeared when a steward gathered up each and every tag and replaced them with a single one, "Madame Oxaloacetic Acid." It wasn't long before the Zoot Suit was anxious to renew his search for the dancing duo, so I passed this letter on to him. I felt the responsibility of not leaving you suspended in a state of limbo but find it necessary to inform you that...

Everything we had between us is altered, as I am altered. Conjugate with another.

Your former Sucrose partner,
Gladys

Mr. Sucrose, disappointed at the lack of leads and my office's inability to solve the disappearance of his wife, has decided to investigate for himself. Scheduled to enter into the training program offered by the Kombucha Factory, he was last seen entering the factory.

INVESTIGATION INTO THE DISAPPEARANCE OF FRANK SUCROSE

To whom it may concern:

I am writing this letter in the hopes that it will provide clues to the disappearance of my dear wife Gladys. I am convinced that we are destined to never see each other again as my journey here within the Kombucha Factory seems to have lead me to a parallel world of her experience.

My transformation at the hands of the factory image consultants was fast and frightening. My personal advisor, Mr. Hex O. Kinase wielded his shears with a terrifying swiftness which caused me to lean heavily to one side. Unfazed, he then introduced me to a gentleman who, I was informed, would support me. I was disturbed to discover that I had been altered to the state of requiring another's support but moving without his assistance proved futile. Begrudgingly, I accepted my new partner's assitance. Taking a closer look at the one who now aided me, I suddenly realized that my support proved to be a phosphate brother identical to the image of the radicals Gladys had so crudely portrayed in her letter. Before I could recover from the shock of this surprise, the two of us were approved and given a name tag which read "Frank 6 Phosphate." My new partner and I were hustled through the make-over room's swinging doors.

We found ourselves on the platform of what I thought was a train station. Above my head was a sign which read "Welcome to the Bacteria Division - The Fermentation Station." But rather than a train station, I found we were in a bus station instead. "Who's in charge?" I shouted but received no response.

Suddenly, three flashing neon signs illuminated the station. Three gates under those signs opened and immediately we found ourselves being pushed and jostled about. With my companion apparently attached to my hip and moving with me as one unit, I began shoving my way back through the crowd. I became more frantic, insisting that there was some kind of mistake, that I wished to ride the train called THE ENERGY EXPRESS. But the tide of other molecules around me was so strong that we were swept up and funnelled through one of the gates. The line of molecules ahead of us led to a rough, idling aluminum-sided bus. As we boarded I strained to catch the eye of the driver as we filed past, hoping to plea for a different mode of transportation. Imagine my shock when he appeared to be the one and the same conjurer Gladys had written about. He was short and stocky with a merry twinkle about his eye. He bade us all a pleasant afternoon as we boarded the vehicle. I managed to peek at his jacket and found the name "Kinase N. Zyme" stitched on it. I reasoned that fate had intervened and placed me on this bus with someone who is probably at least related to Gladys' fellow traveler. When the bus was loaded, he closed the doors and, with a blast of the horn and the grinding of gears, our bus lurched onto a stretch of blacktop labeled "EMDEN-MEYER-HOFF GLYCOLYTIC THOROUGHFARE."

At last able to catch my breath, I turned to my recently acquired assistant and smiled. He was everything she had described. What she thought was the manly sort, I felt was a bit rugged and very simple in expression, lacking the sophisticated refinement of a gentleman. I sat pensive for a moment dwelling on my Gladys' uncharacteristic response to someone so simple and raw. His rough burly hand was thrust in my direction by way of introduction, but I thought I would test my theory of who he was. "No introduction necessary. You're Phosphate, right?"

"Why, yes. But, how did you know?" a look of genuine surprise flooded his face.

"Long story," I replied. I wasn't going to reveal myself. His apparent naive innocence was not going to cause me to spoil my cover. I hoped instead to use his lack of sophistication against him in order to induce him to tell me all that he might know about Gladys.

Suddenly, we were all thrown to the left as the bus pulled off the highway. A trio of individuals milled about on the side of the road. There was a set of phosphate triplets. Their hand-scripted armbands displayed the letters "ATP." I watched through the windows as the trio man-handled one another until one jumped onto the bus. The remaining two brothers jeered at their departing brother, ripped off their ATP armbands and threw off their shirts to expose "ADP" (adenosine diphosphate) written across their T-shirts and, arm in arm, the pair trudged down the road, their heads hanging low.

The driver closed the door and steered it back onto the road.

I had risen to my feet and was standing in the aisle as the new single phosphate moved toward me. He gave me a penetrating look, then asked me where I was sitting.

"Right here." I said, pointing to my seat.

"Good. Looks like I'll join *you*."

He sighed heavily. "Such a silly fight. Over such minor philosophical differences."

"Oh?" I queried.

"Yeah. They felt that we shouldn't bring any outsiders into our organization, but I feel that diversity is essential to our success."

"I quite agree," my original phosphate partner replied.

"You do?" Phosphate's face lit up.

"Oh, very much so. Would you like to join us?"

How amazing! I thought. This undercover stuff is a piece of cake. An opportunity to infiltrate the ATP group and improve my ability to find out what happened to my ill-fated partner has landed in my lap this quickly. I'm good, I thought. When this is all over, I think I'll look into this detective business as a way of making a living.

"Oh, we would consider it an honor," I responded, a wide self-satisfied grin spanning my face.

I looked up toward the front of the bus and found the driver's eyes staring at me through his mirror. He winked and smiled. I quickly lowered my eyes distressed by the driver's actions.

"I see by your name tag that your first name is Frank. How 'bout us becoming "Frank Di Phosphate?" (fructose 1,6 diphosphate), our new phosphate friend asked.

"That pretty much covers the bases, don't you think?" my original phosphate joined in.

"Of course." I responded. And my original phosphate and I

ripped our name tag from our chest and threw it to the floor. All three of us together shared rowdy handshakes and boisterous back clappings.

Our joviality was cut short when suddenly the driver barked, "Everyone but Frank Di Phosphate gets off the bus here!" as he slammed the brakes on and flung open the bus door. I looked out the windshield and discovered that we were at an uninviting, desolate stretch of the thoroughfare. The roadside was littered with pale and withered vegetation. A closer look revealed a run-down gas station. Old-fashioned hand pumps, decrepit with age, their hoses cracked and split, leaned precariously out of the concrete island. A rickety, peeling, wind-scoured sign bounced in the breeze. The cracking, faded words "Dihydroxy Acetone" were barely discernible across its surface.

"Hurry up, Chop, Chop. *Andales*, you noxious characters," the driver bellowed.

"Whew." I couldn't help noticing the truly, sickeningly sweet, acrid smell of the group as they flowed past me. My new-found comrade however was incensed at the driver's treatment of the disembarking passengers. "This is outrageous! What have these poor souls done to warrant such degrading actions on your part. I insist that you refrain from this injustice!" he shouted, as he jumped from our seat and started down the aisle toward the bus driver.

"Oh, yeah?" The chubby little bus driver cockily inquired.

"Yeah!" the first phosphate responded, nearly bending over double in order to defiantly shove his face into the driver's. Without a sign of fear and not missing a beat, the driver launched into action. He flew from his seat backing the mountain-of-a-phosphate into the empty front seat.

Straining his neck to look up at the antagonistic radical, he replied, "Well, sonny boy, my response to you is ... Why don't you go join them! In fact, that works out just fine. Now we'll have 'Dihydroxy acetone *phosphate*.' Put that in your pipe and

106

smoke it!" Surprisingly strong for his size, the driver grabbed phosphate by his shirt and propelled him over the rail and out the door. Spinning around, he slammed the door shut, jumped into his seat, reengaged the bus gears and spinning gravel from the tires, we sped away from the scene. I could hear the rhythm of a chant rising in the distance, "Acetone, Acetone, proud to be a Ketone."

"Yow! Isn't that serious?" I asked my remaining phosphate.

"Yeah, I'd say. I wouldn't want to be in that body," he replied. He gave me a strange look and began backing away from me.

"What? What is it?"

"Ahh.... you've changed. It must have happened when all those smelly passengers passed by you. It's OK. I still recognize you, but you're gonna' have to take another name."

"Really? I did notice I didn't feel very well when they went by, but I feel fine now. Are you sure I need a new name?"

"Yeah, I'm sure. This time I think 'Glen 3 Phosphate' (glyceraldehyde 3-phosphate) works."

I decided that the time was right to make a tentative inquiry concerning Gladys. I leaned forward to position myself closer to my companion. "Excuse me, my friend, but I would like to..."

I stopped short and raised my nose to determine the source of stinking cigar smoke. I was surprised to discover that we had stopped again and there, standing on the bus steps, was none other than Nic. O. Tinamide A.D. (nicotinamide adenosine dinucleotide). There was no doubt. The rude waiter appeared just as Gladys had described him in her letter, corpulent, coarse and pungent. I stared wide-eyed as he turned his gaze in my direction.

"Hey, the boys ready, N. Zyme?" he asked the driver as he continued to stare at me.

"Yep," the driver replied, "They're back there. I"ll go fetch 'em."

The driver rose and ambled down the aisle toward us. He

stopped at our seat, reached under it and surprised me by extracting two screeching flapping cockatiels (2 hydrogen electrons) from nowhere.

"Wow," I shouted, "Where did they come from?"

"Uh, duh... What kind of dweeb do you got there, N. Zyme?" the roly-poly waiter uttered contemptuously while removing his cigar from his dank tobacco-stained lips.

"Hey, Hey," I sputtered.

"Having trouble spitting it out, buddy? I got more verbal articulation from some skirt on THE ENERGY EXPRESS."

"What? Skirt! Why you!" I rose to my feet. My face turned crimson as my temples pounded.

"I rest my case." He turned toward the driver, "Come, my lovelies, it's off to the shuttle for you." He took the screaming birds and placed them under a covered dish he held in his hand.

"Wait, wait!" I cried.

"Later N.," Nic said to the driver, then stepped off the bus. "Hey you," he yelled back to me, "maybe next time we meet, you'll be able to converse in more than single syllables." He placed the silver dish upon a linen-covered cart and shoved it down the road. The bus began moving down the blacktop. I jumped to my feet and yelled for the driver to stop.

He shook his head. "No, siree. I only make the specified stops that are on my orders," he explained as he pointed to a paper-filled clipboard hanging from his dashboard." There is only so much time to get through this cytoplasm."

"Please, you don't understand," I implored.

He drove on, ignoring my pleas. Despairingly, I sunk into the nearest seat. My friend, Phosphate, rejoined me placing a consoling arm around my shoulder.

Suddenly, in the middle of the road, stood our former partner, Phosphate. The bus pulled up and, while it was still rolling, he jumped onboard. He launched into an explanation before we had opportunity to ask for one.

"They were ungrateful. They just drifted away like a cloud, abandoning me."

"Oh good. Now we can be 'Agents 1,3 and Di'!" (1,3 Diphosphoglyceric acid), my phosphate partner declared, a broad grin spanning his face. He was obviously pleased with himself for these new names. "Cool, huh?"

"Wait a minute! Look there. It's my brothers," my original phosphate friend cried, jumping to his feet.

"Right on time," Driver N. Zyme muttered.

"Time for what?" I asked.

"Oh, just time..." was his response as he pulled the bus off to the side of the road.

"Wow! Hey buds," one of my phosphate companions yelled. His brothers leapt onto the bus and embraced their sibling enthusiastically. Their bonding was so immediate and complete that before I realized it, they had once again donned their ATP armbands and jumped out the bus door.

"Hey, what about me?" I shouted.

"You have the other phosphate. You two can pull it together. And by the way, you should change your name to just 'Agent 2'," (2-Phosphoglyceric acid) he shouted back at me as the bus once again returned to the road.

Stunned, my mind began to tumble. I'm becoming an acid. I shuddered at this realization. That's twice that I have been referred to as an acid and I'm getting more acidic. I don't like this. My thoughts were interrupted by the feeling of being watched. I looked up and, just as I expected, N. Zyme was looking at me in his overhead mirror. He smiled and winked. Phosphate, my friend, squealed and jumped back.

"Yow! You're sharp!"

"Sharp? What do you mean by 'sharp'?" My voice had become high pitched and strained. I looked down and realized something drastic was happening to me.

"Why, you've changed once again and, this time, you're a

pyruvic."

"A pyruvic? How can that be? That's what my Gladys had become for a while. How is it possible?"

"Look, I don't know who Gladys is, but this is scary. You're now a phosphoenol pyruvic acid."

"Oh my, oh my," I began to lament as I rose to pace the aisle of the bus. "How is this happening to me?"

It was at that moment I realized who the real culprit was. Both mine and my dear Gladys'. I spun around to face my demon when a shout halted my advance.

"Look there! Just ahead! My brothers!" There standing in the middle of the road were the estranged phosphate twins. The bus pulled up. My last hope leapt from the bus. And the same reunion scene was reenacted, restoring the triplet powerhouses. Replacing their ATP armbands, they moved off, disappearing into the surrounding countryside.

"I know the truth! I know you're to blame," I seethed as I moved to the front of the bus.

I was taken aback at the heightened pitch of my voice and the lack of strength I was experiencing. I collapsed into a nearby seat.

The driver never looked up as he continued driving.

"Answer me! You're the one, aren't you?" I muttered as I struggled to raise myself.

He did not respond. I pulled myself along one seat at a time. He began to whistle a show tune.

Just then a fork in the road caused the driver to halt the bus. I struggled to continue my agonizing trek down the aisle. I was forced to rest upon a nearby seat. I looked through the windshield and saw a huge road sign.

The driver reached down and removed the clipboard from its hook on the dash. As he flipped through the papers, his whistling was beginning to drive me crazy.

"Tell me! I beg you. You did it to her and now you've done

it to me. Aren't I right, **N. *Zyme***?"

"Probably, but if it's any consolation, it wasn't me personally. There are thousands of us all over the place doing what has to be done." He responded to my anguish so casually it was maddening.

Finally he looked up, still smiling, with a glint in his eye. "Hey, I'm only doing my job, and now that *you* are Pyruvic Acid, I have to determine where you are to go from here."

I was devastated and forlorn for I was drained of the ability to fight him.

I turned my attention to the sign outside. To the right was printed **"Aerobic Pathway** to Track 2 of the ENERGY EXPRESS". To the left was printed **"Anaerobic Pathway** to Carbon Dioxide and Lactic Acid, Acetic Acid or Ethanol."

Joy flooded over me as I realized I had one more opportunity to locate my darling Gladys. "Take me to the ENERGY EXPRESS. I know someone who waits for me there," I pleaded.

"Well, I'm afraid not. According to my orders from the Kombucha Factory, you're destined to become carbon dioxide and vinegar. It's a done deal, buddy. Hang on."

With that he shifted gears and headed down the dimly lit **Anaerobic Pathway**.

And so ended the letter.

This unsigned letter was discovered floating outside the drainage pipes of the Kombucha Factory. Upon subsequent investigation no evidence was found linking the factory to the disappearances, although there have been increasing reports of missing molecules, all with the same names, Frank and Gladys Sucrose. The mystery continues...

Detective Findem rips the last page from the typewriter's rollers and scribbles his name. Rising slowly, he dons his threadbare trench coat, lifts his beaten fidora off the coat hook and places it squarely upon his balding head. Then, after pulling up his collar to hug his fleshy jowls, he exits his office. He trudges across the street to the gates of the Kombucha Factory, ambling up to the office door. The head of the make-over division, Mr. N. Zyme, greets him with exaggerated warmth. The detective shrugs off the manager's clammy hands from his shoulders. "Here. This is the report you requested concerning the disappearance."

"And?" queries the manager expectantly.

"The page on top, Mr. N. Zyme, is my resignation."

"But why?" N. Zyme asks, "are the benefits not to your liking?"

"No, Mr. N. Zyme, the benefits are not the problem. The problem is that I smell a rat."

With that the detective turns on his heels and exits.

"Good, another off the trail." Mr. N. Zyme snickers as he moves to the phone book. He flips to the private detective section and checks off yet another name. "That takes care of the F's, now on to the G's. Ahh... Investigator Gonn A. Getum," he chuckles, as he lifts the phone receiver.

⟞ THE END ⟝

~~ PART 4 ~~
Glossary And Loose Ends

GLOSSARY

acetone—a substance with a fruity, ethereal but acrid odor which is produced in an anaerobic environment as a byproduct of fermentation of the Kombucha beverage. This substance can cause damage to the kidneys.

acetylcholine (ACh)—this substance plays an important role in the transmission of nerve impulses.

acetic acid—the medical and industrial term for vinegar.

acetyl SCoenzyme A—a condensation product of coenzyme A and acetic acid.

acid—any substance which contributes a hydrogen ion into solution. It is the liberation of this ion that will cause a solution to go more to the acid pH (0-6.9). Acids also reduce or neutralize alkalinity.

acidosis—a condition of excessive acidity of body fluids due to an accumulation of acids or the excessive loss of alkali buffers (bicarbonates).

adenosine diphosphate (ADP)—breaking this word down, we come up with a nucleotide containing adenine, ribose and 2 phosphoric acid molecules. ADP is readily rephosphorylated (a phosphoric acid group added) to become ATP, the principal unit of energy storage.

adrenal acidic—This term will not be found in the *Merck Manual* (the diagnostic guide book for physicians) although it is a syndrome which, from my experience, precipitates serious conditions. Persons who experience unresolved stress are subject to the adrenal secretions (epinephrine and aldosterone) being released over prolonged periods of time. These secretions will severely alter the individual's biochemical balances. If

113

these aberrant secretions go unchecked, an individual will produce too much of the enzyme cholinesterase. This cholinesterase (ChE) will increase cell production of acetic acid by stimulating cell metabolism and can change the blood pH values by increasing acidity from .35 to 1.4 points. This process destroys acetylcholine. When acetylcholine is not available, certain nerve plexi are affected, including all preganglionic, all postganglionic parasympathetic fibers and postganglionic sympathetic fibers to the sweat glands and efferent fibers to the skeletal muscle. It can cause sweaty palms, spontaneous sweating, shortness of breath, nervous stomach or nausea, and chemical impulses to set the heart racing. These symptoms often culminate in anxiety attacks. This biochemical mess will affect virtually every system of the body from digestion to respiration.

aerobic—a process or condition which occurs in the presence of oxygen.

aldosterone—a steroid hormone, the most active mineralocorticoid secreted by the cortex of the adrenal glands: regulates sodium, potassium, and chloride (electrolytes) metabolism via the kidneys. It will cause the body to hold or release water in order to maintain proper electrolyte levels.

alkaline—a solution or condition in which there are a greater ratio of OH^- molecules to H^+ molecules present. Values range from 7.1 to 14 on the pH scale.

alkaloid—one of a group of organic alkaline substances obtained from plants. Alkaloids react with acids to form salts.

amino acids—a compound containing both an amine group (NH_2) and a carboxyl group (COOH). There are two types, essential aminos which are not synthesized in the body and nonessential ones which are.

aminophylline—a combination of theophylline and ethylene diamine used as a drug in the treatment of breathing disorders, asthma, bronchitis and emphysema.

anaerobic—a condition which exists or a process which occurs without the presence of oxygen.

adenosine triphosphate (ATP)—the major source of cellular energy found in all cells. It is composed of a nitrogenous base (adenine), ribose sugar and 3 phosphoric acid radicals. It is the chemical removal of the phosphoric radical that releases energy.

autonomic nervous system—refers to the involuntary or self-regulating portion of the peripheral nervous system. This system runs all the functions that we do not have to conciously direct such as: heartbeat, breathing and digestion, reflexes etc.

bacteria—the simplest type of single cell organisms. There are three principal forms of bacteria. For the Kombucha culture, we are concerned with the rod-shaped bacteria known as bacilli. Most bacteria are relatively constant in form in growing cultures. In old cultures or cultures grown under adverse environments, aberrant forms such as oversized or Y-shaped occur. These are considered by some microbiologists to be involuting or degenerating forms of these bacteria.

bicarbonates—any salt containing the HCO_3 (bicarbonate) anion. Anions are negatively charged ions.

blood buffers—a chemical system composed of bicarbonate ions and dissolved carbon dioxide. This system is regulated by the kidneys and respiratory system and maintains the pH of the blood near 7.4.

butyric acid—a fatty acid that is found in butter and cod liver oil. It has been reported to have a protective benefit to cellular membrane. In humans, it is most commonly encountered as the cause for rancid-smelling perspiration. Butyric acid combines with bacteria on the body to produce body odor. Its industrial uses include disinfectants and emulsifying agents.

caffeine (1,3,7-trimethyl-3,7, dihydro-1H purine-2,6, dione)—a xanthine which is further degraded to purines, then to uric acid. It has a stimulating effect on the central nervous system, mainly affecting the cerebrum. It has a diuretic effect on the kidneys, stimulates striated voluntary musculature and has a number of effects on the cardiovascular system.

candidiasis—the unhealthy proliferation of the naturally occuring *Candida albicans* yeast within the human body. This condition is provoked by the taking of antibiotics, birth control pills and pharmaceutically derived corticosteroids. Yeast infections of the colon, the vagina and thrush are visible signs of this condition, however, the most drastic consequences arise when the yeast has infiltrated the blood, organs and brain.

cartilage—a form of connective tissue usually with no blood or nerve supply of its own. Healing cartilage is difficult because of this lack of blood supply. Hence the use of such substances as glucosamines and Chondroitin Sulfate A (extracted from beef sources) to replace damaged cartilage cells in humans as a treatment for certain degenerative arthritic conditions and joint, tendon and ligament injuries.

chelation—a routine natural process that is necessary for many enzyme functions by wedding two unrelated substances, organic and inorganic, into a compatible working partnership. Chlorophyll, the green pigment of plants is a chelate of magnesium and its cousin, hemoglobin, the oxygen-carrying red-colored molecule of blood which itself is a chelate of iron.

Chronic Fatigue Syndrome (CFS)—Originally referred to as "yuppie flu," the syndrome is typified by chronic immunosuppression with flu-like symptoms, extreme fatigue, chronic swollen glands, sensitivities to a vast range of substances both environmental and dietary, arthritis-like body aches and pains and devastating deep, dark depression. This syndrome, from my experience, is precipitated by prolonged unrelenting stress hormone production and the body's ultimate exhaustion phase which makes it vulnerable to otherwise relatively harmless viruses. Initially treated as a mental or attitude problem, practitioners are now scrambling to find a cure for this increasingly more common condition that is now emerging among children as well. The condition has more indicators of being a lifestyle issue and its treatment much more focused on stress prevention and remedial lifestyle practices. This syndrome is the tip of the iceberg of the potential conditions our lifestyles and cultural behaviors will produce if deterioration of the environment and the implosion of society continues.

cirrhosis—a chronic liver condition characterized by the degenerative "hardening" of liver cells.

collagen—a fibrous protein forming the main organic structures of connective tissues.

detoxifier—a substance which removes toxic or poisonous agents from the body.

deoxyribonucleic acid (DNA)—the carrier of genetic information for all organisms except the RNA viruses. DNA is constituted of protein bases (adenine, guanine, cytosine and thymine), a cell sugar (deoxyribose) and phosphoric acid. It is a code for determining hereditary characteristics.

disaccharide—a sugar composed of two simple sugar molecules. Sucrose is a disaccharide containing one molecule of glucose and one molecule of fructose.

edema—swelling due to accumulation of extracellular fluid causing those fluids to be retained outside cells between interstitial tissue structures. Edema is commonly found in the joint tissue.

ethanol—ethyl alcohol or grain alcohol.

eukaryotic cell—a cell which has a true nucleus and membrane-bound organelles in which cellular functions are performed.

extracellular—outside of the cell.

fatty acid—a hydrocarbon in which one of the hydrogen atoms has been replaced by a carboxyl group (COOH). You may be surprised to learn that acetic acid, or vinegar, is by definition a fatty acid.

fermentation—the decomposition of more complicated substances by the action of enzymes or ferments which are combined with oxygen. Bacteria, molds, and yeasts are the principal groups of organisms involved. The biochemical pathway of fermentation is know as glycolysis. Although fermentation is considered an anaerobic process, the breakdown of glucose and acetic acid to citric acid (an aerobic process) is designated as fermentation. Fermentation does not yield much energy; the end products still retain abundant energy in the form of their chemical bonds. In the breakdown of glucose to lactic or pyruvic acid there is a net gain of only 2 molecules of ATP. However, in aerobic respiration (Krebs cycle) in which the glucose is broken down all the way to CO_2 and water, at least 32 molecules of ATP are produced. (see glycolysis)

formic acid—a rather strong acid which occurs naturally in animals and is found in the muscles. It is one of the irritants released in bee stings and ant bites.

free radical—a singlet oxygen (O⁻) which indiscriminately oxidizes or burns off cells. Free radical activity, although beautiful when happening to deciduous trees in autumn, will wreak havoc in the human body. It has been linked to aging, heart disease, arthritis and fatigue. Depriving the body of bioactive oxygen (O_2), free radicals destroy otherwise healthy cells. Think of them as stealthy little "ninjas" armed with plastic explosives and attaching those explosives to any vulnerable tissue site in your body. A few of the ways we gather free radicals into us is from eating fried fats, exposing ourselves to excessive ultraviolet light and by breathing polluted air.

fructose—a fruit sugar, sometimes referred to as levulose. It is a component of sucrose.

fungus—a vegetable cellular organism that subsists on organic matter. Yeasts, mushrooms and molds fall into this broad category. Mushrooms are designated by the formation of a tissue (pseudoparenchymatous tissue) in the fruiting body giving it an "organ-like" structure. Kombucha is better classified as a culture, rather than a mushroom, simply because it does not reproduce via spores, as mushrooms do. Although Kombucha appears to have the pseudoparenchymatous tissue, its mode of replication distinguishes it from mushrooms. It may reveal itself to be a complex organism within its own class.

galactose—from the Greek roots *gala* (milk) and *ose* (sugar), a simple sugar commonly found in the lactose of milk.

ganglion—a mass of nerve bodies outside of the brain and spinal cord.

gluconic acid—results from the process of hydrolysis by which a water molecule is split, leaving an extra H^+ (hydrogen ion) attached to a glucose molecule forming a sugar acid. It is produced by the bacterium gluconicum.

glycolysis—the breakdown of sugar by a hydrolyzing enzyme. In the course of a series of reactions, glucose (a 6-carbon sugar) is broken down to two molecules of pyruvic acid, a 3-carbon compound. The pyruvic acid then may be reduced to form lactic acid, or it may lose carbon dioxide to form acetaldehyde, which is subsequently reduced in the Krebs cycle to carbon dioxide and water.

glucoronic acid—an oxidation product of glucose. As one of the body's most potent liver detoxifiers, it is especially helpful in removing toxins of the phenol groups which include pesticides, herbicides and plastics.

glucosamine—a constituent of mucopolysaccharide.

glucose—from the Greek root *gluc* (grape). A 6-carbon monosaccharide which is used by the body for energy production within the cell. In the human body, glucose requires insulin to carry it past the cellular gates into the cell.

H^+—a symbol for a positively charged hydrogen ion.

hyaluronic acid—a mucopolysaccharide acid found in the connective tissue which acts as a binding, lubricating and protective agent. It is also found in synovial or joint fluid and in the vitreous humor (the clear watery fluid which fills the inner space of the eyeball).

hydrolyze—to cause to undergo hydrolysis. Hydrolysis is any reaction in which water is one of the participants. Generally, the H_2O molecule is split into its components OH^- and H^+ and these individual units are attached to the free end of another molecule.

inorganic—a substance that does not contain hydrocarbon in its chemical structure.

ion—an atom or group of atoms (radical) carrying an electrical charge.

ketone bodies—a group of compounds produced during the oxidation of fatty acids. These include acetone, acetoacetic acid and beta-hydroxy-butyric acid.

Krebs cycle—also referred to as the citric acid cycle. This vital process is a complicated series of chemical reactions in the body involving the oxidative metabolism of pyruvic acid and then converted into chemical form to synthesize ATP.

lactic acid—an acid created in the fermentation process of certain microorganisms and, in humans, in the muscles during activity. It is produced by the process of glycolysis.

malic acid—from the Latin root *malum* (apple), this acid is found in large quantities in apples, cherries, apricots and other sour fruits. It is said to be an oxygen carrier and in that capacity seems to cause the liver viscera to relax aiding the detox process. Another possiblity for its potential therapeutic benefit is in its transformation into maleic acid. A water molecule removed from malic acid forms maleic acid, a cis-fatty acid. Cis-fatty acids are known to carry unhealthy fats out of the body and reduce the impact of trans-fatty acids.

medium—a substance used for the cultivation of microorganisms. In the case of the Kombucha beverage, the tea and sugar solution is the medium.

metabolism—the sum total of all physical and chemical reactions in cells and in the body for the maintenance of life.

metabolite—any product of metabolism.

mitochondria—a cell organelle which is responsible for energy production. It is called the "power house" of the cell. An organelle is an "organ" of a cell. This organelle has two membranes surrounding it. Inside the first wall are enzymes that break glucose into pyruvic acid. Inside the second membrane, the surface area is convoluted to increase the surface area. In these circuitous hallways are orderly arrays of enzymes that complete the reduction of pyruvic molecules into carbon dioxide and water.

monosaccharide—a simple, 6-carbon sugar. The three most often encountered in human physiology are: glucose, fructose and galactose.

mucopolysaccharide—a thick gelatinous material that is found in many places in the body. It glues cells together and lubricates joints.

myelin—a fatty substance composed mostly of phosphatidyl choline forming a sheath or covering around many nerve fibers.

OH—symbol for the hydroxyl ion molecule composed of one oxygen and one hydrogen atom. The placement of either a $^+$ or $^-$ sign denotes the electrical charge either positive or negative of that molecule or atom.

organic—a substance which contains a carbon compound and pertains to or is derived form animal or vegetable forms of life.

organic acid—an acid containing a carbon compound. Acetic acid (or vinegar) is an example. All the acids in the Kombucha beverage are organic acids.

oxalic acid—a simple dibasic acid, it often functions in biochemistry as a chelating agent. It plays a significant role in the preparation of fructose and glucose for entrance into the Krebs cycle. Its potassium or calcium salt occurs in cranberries, rhubarb, chard, gooseberries, spinach and beet greens. Oxalic acid can bind up calcium making it unabsorbable. Oxalic acid can be detrimental if ingested in large quantities.

oxidation—a reaction of something with oxygen resulting in the loss of electrons. Fire and rust are examples of oxidation.

prokaryotic cell—a bacterium which does not have a true nucleus or mitochondria and reproduces by cell division.

pH—abbreviation of "potential of Hydrogen." It denotes the degrees of acidity or alkalinity of a substance. .1 to 6.9 on the pH scale denotes acid (.1 being the most acid, 6.9 the least). 7 is neutral. 7.1 to 14 denotes alkalinity, (7.1 being the least alkaline and 14, the most).

protein—a complex nitrogenous compound which yields amino acids when hydrolyzed. Proteins are essential to the growth and repair of tissue.

ribonucleic acid (RNA)—composed of nitrogenous bases, ribose sugar and phosphate. There are three types of RNA: transfer RNA, ribosomal RNA and messenger RNA. RNA controls protein synthesis in all living cells and takes the place of DNA in certain viruses.

SCoenzyme A—a factor necessary to metabolism which consists of adenine, ribose, three phosphate radicals and pantothenic acid, as well as the amino acids methionine and cysteine.

sucrose—a disaccharide made up of one glucose molecule and one fructose molecule found in many plants, especially sugar cane and sugar beets.

sugar—a sweet-tasting carbohydrate belonging to the monosaccharide or disaccharide group.

synovial fluid—a colorless, viscid, lubricating fluid in joint cavities, tendon sheaths and bursa. It contains mucin (a glycoprotein, sugar protein), albumin (a group of simple proteins), fat and mineral salts.

tea (for Kombucha purposes)—a beverage made from the dried or treated leaves of the *camilla sinensis* plant. The water soluble substances of tea include: tea tannins, caffeine, protein bodies and gummy matter. The average caffeine content of black tea is about 3% total volume, averaging about 40 mg per cup.

theobromine—a compound found in trace amounts in tea leaves. Its physiological properties are similar to those of caffeine as a diuretic, smooth muscle relaxant, vasodilator and heart muscle stimulant.

theophylline—a compound also found in small quantities in tea leaves, it is a smooth muscle relaxant. As a muscle relaxant, theophylline is used as the active ingredient in bronchodialators for the treatment of emphysema, bronchial asthma and acute bronchitis. It also stimulates the heart muscle as a coronary vasodilator and is a diuretic and respiratory center stimulant.

trace minerals—minerals that are found, supplied or used in very minute quantities but essential to the body's function.

trans-fatty acids—made by exposing natural cis-fatty acids to heat or hydrogenation as in margarine and fried foods. Trans-fatty acids are rigid artificial molecules which cause the blood to thicken, become sticky and accumulate along arterial walls.

usnic acid—an antibiotic sometimes purported to be a component of the Kombucha beverage. Usnic acid, however, is derived from usnea lichen and as the culture is not a lichen, the possibility of usnic acid occurring in the culture is very unlikely.

vasodilator—a substance which causes the dilation of a blood vessel, thereby increasing blood flow.

xanthine—a substance which possesses stimulant properties affecting muscle tissue, especially the heart muscle. Xanthines are generally broken down to form other purines and eventually uric acid.

yeast—a broad category of single-celled, usually rounded fungi that reproduce by budding and sometimes cell division. Saccharomyces cerevisiae are the most commonly used, especially in brewing beer, making alcoholic liquors and in baking.

RESOURCES

Here are some sources for the organizations and equipment mentioned in this book. We are offering through our mail-order department aquarium heaters and pH strips. These items can be hard to find, especially if you live in a non-urban area. Please use the order form in the back of this book for your convenience.

The Kombucha Foundation—a data bank for culture "needers" and "breeders"
P.O. Box 661056
Los Angeles, CA 90066

Shepard's Garden Seeds—for seed propagation mats
Order Dept. call (203) 482-3638
Item # 7125 - 21" x 15 1/4" x 1 1/2"
Item # 7218 - 36" x 17" x 1 1/2"

ABOUT THE AUTHOR(S)

Alana Pascal is a Certified Clinical Nutritionist with a strong background in biochemistry and 23 years of experience in the alternative health field. She has a special interest in Anthropological Nutrition and is also an historian. She designed and taught a state certification program in clinical nutrition at Heartwood Institute in Northern California from 1982 to1987. In 1987 she moved to Los Angeles with her two children, Sierra and Caleb, where she is a consultant to clinics, businesses, corporations and individuals.

Lynne Van der Kar is a Certified Massage Therapist born and raised in the Los Angeles area. She was an artist and potter for over 15 years. In 1988, she became interested in the Healing Arts and alternative health care. She continues her studies with many teachers and healers from around the world learning various forms of bodywork and massage. She is recommended by the The Esalen Institute in Big Sur, California, and today has a private practice in Malibu, where she and her beloved late husband built their home.

BIBLIOGRAPHY

ANON.: Cancer protection: Regular or decaf?
Science News Vol: 146 Iss: 4 Date: Jul 23, 1994 p: 61

ANON.: Green Tea for your Medicine Cabinet. Better Nutrition
for Today's Living Vol: 56 Iss: 9 Date: Sep 1994 p: 38-41

BLAKESLEE, S.: Does Ordinary Baker's Yeast Hold Secret to
Curing Common Cold?, New York Times, Nov. 22, 1994

BODANIS, D.: The Secret Garden. N.Y., 1992

BOHINSKI, R. C.: Modern Concepts in Biochemistry. Boston, 1973

BRAGG, P. and P.: Apple Cider Vinegar Health System.
Santa Barbara, 1993

CHAITOW, L.: Amino Acids in Therapy. N.Y., 1985

DARNELL, J., LODISH, H., and BALTIMORE, D.:
Molecular Cell Biology. N.Y., 1990

DHARMANANDA, S.: Chinese Herbology. Portland, OR., 1989

DORLANDS ILLUSTRATED MEDICAL DICTIONARY
26th ed.: Phila., 1986

ENCYCLOPEDIA AMERICANA. Danbury, CT., 1993

FASCHING, R.: Tea Fungus Kombucha. 4th ed.: Steyr, Austria, 1994

FRANK, G.W.: Kombucha. Steyr, Austria, 1994

JARVIS, D. C.: Folk Medicine In Vermont. Greenwich, CT., 1958

KAPTCHUK, T. J.: The Web That Has No Weaver. N.Y., 1983

KIRSCHMANN J. D, and DUNNE, L. J.:
Nutrition Almanac. N.Y., 1984

MERCK MANUAL 14th ed.: Rahway, N.J., 1982

SHALLECK, J.: Tea., N.Y., 1972

TABOR'S CYCLOPEDIC MEDICAL DICTIONARY
14th ed.: Phila., 1982

INDEX

honey 61
hyaluronic acid 32
hypoglycemia 36

immunosuppression 69
inositol 25
insulin 23, 27, 59

Jarvis, Dr. Henry 32

ketones 36, 54, 80
kidneys 20, 32, 39, 40, 41, 54,
64, 80
Krebs cycle 18, 85
Kvass 14

lactic acid 14, 15, 30, 31, 85, 86
lactobacilli 12, 14, 30, 31, 69, 82
large intestine 39
ligaments 45, 68
liver 20, 21, 26, 30, 31, 32, 40,
43-48, 65, 68, 69, 70, 76, 77, 82

malic acid 18, 19, 40, 85
Manchurian mushroom 13
maple syrup 61
menopause 82
methionine 21, 26, 71
minerals 27, 28, 42, 44, 56, 61
mitochondria 85
mucopolysaccharides 32
muscle testing 72

nerves 22
niacin 23
nucleic acid 25, 27, 28
nutritional yeast 20-21, 67

oxalic acid 18, 85
oxygen 24, 51, 52, 54, 66, 68, 81

PABA-para-amino benzoic acid 25
pancreas 38
Pauling, Dr. Linus 28
pH 33, 35, 38-41, 61, 80, 81, 83
acid 18, 33, 56, 58, 71, 84
alkaline 33, 34, 35-36
pH strips 50

phenol 32
picorna rhino virus 67
PMS 82
pregnancy 71
prokaryotic 85
psoriasis 46
purines 65

RNA 21, 27, 67

saccharomyces yeast 12, 15, 17,
20, 66, 74
SCoenzyme A 26
sucrose 17, 61
sugar 16, 17, 31, 44, 49, 58,
61, 75

tea 14, 16, 49
tea, black 49, 50, 62, 63, 65
tea, green 50, 62, 65
tea, herb 62
theobromine 64
theophylline 64, 69
thyroxin 23, 44, 68

uric acid 65
usnic acid 58

vinegar 15, 29, 32, 43, 49, 51, 78
vitamin C 21, 28

xanthine 65

yeast 11, 15, 16, 17, 18, 27, 28,
40, 50, 52, 58, 61, 67, 70, 75,
76, 78, 79

ORDER FORM

To brew Kombucha to perfection, we are offering these hard-to-find products to assist you:

The aquarium heater is manufactured by Instant Oceans - 200 watt - measuring 11 inches in length. Just set the temperature gauge to the recommended 72 to 80 degrees F. and the automatic thermostat does the rest.

The easy-to-use pH paper strips measure the entire pH scale from 0 to14. There are 100 color-coded strips in each box with a color chart to tell the exact pH.

QUANTITY	TOTAL
_____ Aquarium Heater @ $35.00 each	$ _____
_____ pH strips @ $15.00 (Box of 100)	$ _____
California residents add 8.25% Sales Tax	$ _____

8.25% of $35.00 = $2.89
8.25% of $15.00 = $1.24

Shipping/Handling Charges:

1 Aquarium heater and 1 box of pH strips - $6.00	$ _____
1 box of pH strips - $1.00 (add $.50 each additional box sent to the same address)	$ _____
TOTAL DUE	$ _____

All orders must be pre-paid. Send check or money order to:

The Van der Kar Press
ATTN: Mail-Order Dept.
P.O. Box 189
Malibu, CA 90265-2855